THE GENUS
JASMINUM
IN CULTIVATION

A BOTANICAL MAGAZINE MONOGRAPH

THE GENUS
JASMINUM
IN CULTIVATION

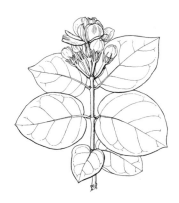

by

Peter Green and Diana Miller

With a chapter on jasmin in the perfume industry by Robert Calkin

Edited by Martyn Rix

Kew Publishing
Royal Botanic Gardens, Kew

Kew

PLANTS PEOPLE
POSSIBILITIES

First published in 2009 by
Royal Botanic Gardens, Kew
Richmond, Surrey, TW9 3AB, UK
www.kew.org

ISBN 978-1-84246-011-5

British Library Cataloguing in Publication Data
A catalogue record for this book is available from the British Library

Production Editor: Michelle Payne
Typesetting and page layout: Christine Beard
Design by Publishing, Design and Photography,
Royal Botanic Gardens, Kew

Front cover illustration: *Jasminum polyanthum* by Lilian Snelling, for *Curtis' Botanical Magazine* t. 9545.

Back cover and flap illustrations: *Jasminum floridum* by Anne Barnard, *Jasminum bignoniaceum* by Dumphy/Govindoo and *Jasminum humile* forma *farreri* by Lilian Snelling.

Frontispiece: *Jasminum multipartitum* by Christabel King.

Printed and bound in the United Kingdon by Henry Ling Limited

Mixed Sources
Product group from well-managed
forests and other controlled sources
www.fsc.org Cert no. SA-COC-001860
© 1996 Forest Stewardship Council

FSC

The paper used in this book contains material sourced from responsibly managed and sustainable commercial forests, certified in accordance with the FSC (Forestry Stewardship Council).

For information or to purchase all Kew titles please visit
www.kewbooks.com or email publishing@kew.org

All proceeds go to support Kew's work in saving the world's plants for life

CONTENTS

LIST OF PAINTINGS

PREFACE

EDITOR'S NOTE

Jasmines are found wild throughout the warmer parts of the world, with the exception of the Americas. Of around four hundred and fifty species in the genus, only twenty-five or so are cultivated in the British Isles; the white-flowered species are grown primarily for their scent, the yellow ones for their flowers, in the case of the familiar *Jasminum nudiflorum* and of *J. mesneyi*, producing the brightest flowers of winter. Several other species are grown commonly in warmer climates in America, Asia and Australia.

This book is based on the studies of Peter Green, who has published many papers on *Jasminum*, mainly during his career in the herbarium at Kew. Peter came to Kew from the Arnold Arboretum and was Keeper of the herbarium, deputy Director of Kew and for many years editor of *Kew Bulletin*. Until 2007, he continued to be a daily visitor to the herbarium, completing various aspects of his work on the Oleaceae.

Peter Green's first jasmine paper was published in 1961, and a series followed, until "Studies in the genus *Jasminum*, XVIII (Oleaceae)", published in 2003.

Diana Miller, who worked for many years as curator of the herbarium at Wisley, has written this book and used much of Peter Green's work for the botanical section, which concentrates on the species of *Jasminum* known to be in cultivation. Peter checked the typescripts of the species as they were completed.

Robert Calkin has kindly provided a chapter of jasmin in perfumery, a valuable oil and an important ingredient of several famous scents, such as JOY by Patou, CHANEL NO. 5 and Dior's EAU SAUVAGE. The subtropical *Jasminum grandiflorum* is the species most often grown for scent production; the tropical *J. auriculatum* and *J. sambac* are less valuable, and the flowers of *J. sambac* are the common ingredient of jasmine-scented tea.

Apart from studies in the herbarium, much of this book, and most of the illustrations are based on Jasmines in cultivation. Guy Sissons of The Plantsman Nursery has introduced new species to cultivation in Europe, particularly from Thailand, and has been most generous in supplying material for study. Geoffrey Herklots prepared most of the line drawings during his work as a botanist in many parts of the world.

Publication of this book was supported by Valerie Finnis, who gave a generous grant to Kew to be used for the production of books connected with *Curtis's Botanical Magazine*, and it is sad that she has not seen this book appear.

ACKNOWLEDGEMENTS

Our thanks are due to the late Valerie Finnis, who gave a substantial grant towards the completion of this book.

We should also like to thank the following for their help:

Robert Calkin, for the chapter on jasmines in perfumery.

The staff of Lindley Libraries of the Royal Horticultural Society at Wisley and in London.

The Botanical and Plant Records staff at Royal Horticultural Society at Wisley.

The Library and Herbarium staff of Royal Botanic Gardens, Kew.

Guy Sissons of Plantsman Nursery, Libourne, Gironde, France; Stella Boswell, née Herklots, for permission to use her father's drawings; George Staples formally of Bishop Museum, Honolulu, Hawai'i; Rodger Elliot of Heathmount, Victoria, Australia; Dr Srisanga of Queen Sirikit Botanic Garden, Thailand; Rod & Rachel Saunders of Silverhill Seeds, South Africa; Bleddyn Wynn-Jones of Crûg Farm Plants; Peter Catt, Liss Forest Nurseries for information on propagation.

And many botanical and horticultural friends for general discussions and advice.

1. INTRODUCTION

THE GENUS JASMINUM AND OVERVIEW OF THE FAMILY OLEACEAE

Jasmines, members of the botanical genus *Jasminum,* are widely grown and much loved for their sweet-scented flowers. Although the genus contains some 200 species, jasmine's reputation as a popular garden plant rests on only a few. All the species are native to the Old World, and when found growing wild in the New World, they are naturalised garden plants that have escaped from cultivation.

Jasminum belongs to the olive family *Oleaceae*, which contains around 28 genera and 900 species of trees, shrubs and climbers, found throughout the world, but especially in south-east Asia. The family includes several important ornamental genera such as *Abeliophyllum*, *Chionanthus*, *Forsythia*, *Fraxinus*, *Ligustrum*, *Osmanthus*, *Phillyrea* and *Syringa*. The olive itself, *Olea europaea* is a very important economic plant producing oil, fruits and wood. *Fraxinus*, *Nestegis* and *Notelaea* are important timber-producing genera. All members of the family have opposite leaves (except for Section Alternifolia of *Jasminum*) which may be simple, pinnate or 1–3-foliolate, without stipules. The flowers are regular, usually bisexual, solitary or in a cymose inflorescence. The calyx is small and 4–15 lobed (or absent in *Fraxinus*). The corolla is radially symmetrical, fused at the base, with usually 4 but up to 12, imbricate, convolute or valvate lobes, sometimes showy but often absent in *Fraxinus*. There are usually 2 stamens, sometimes 4, attached to the corolla tube; the anthers dehisce by slits. The ovary is superior, joined with 2 cells with generally 2, sometimes one to many, ovules with axile placentation; the style is terminal, the stigma 2-lobed. The fruit may be a capsule, a berry, a drupe or a samara.

The more common genera of the Oleaceae may be arranged as follows:

    Oleoideae
    Corolla parts 4 (rarely 3 or 6); ovules 2 per cell
        Fraxineae
            Fruit a samara e.g. *Fraxinus*
        Oleeae
            Fruit a drupe or 2-celled capsule e.g. *Chionanthus*, *Ligustrum*, *Olea*, *Osmanthus*, *Phillyrea*, *Syringa*
    Jasminoideae
    Corolla parts 4–12; ovules 1, 4 or many per cell
        Jasmineae
            Fruit a capsule or berry e.g. *Jasminum*, *Menodora*
        Fontanesieae
            Fruit indehiscent e.g. *Fontanesia*
        Forsythieae
            Fruit a tough capsule e.g. *Abeliophyllum*, *Forsythia*

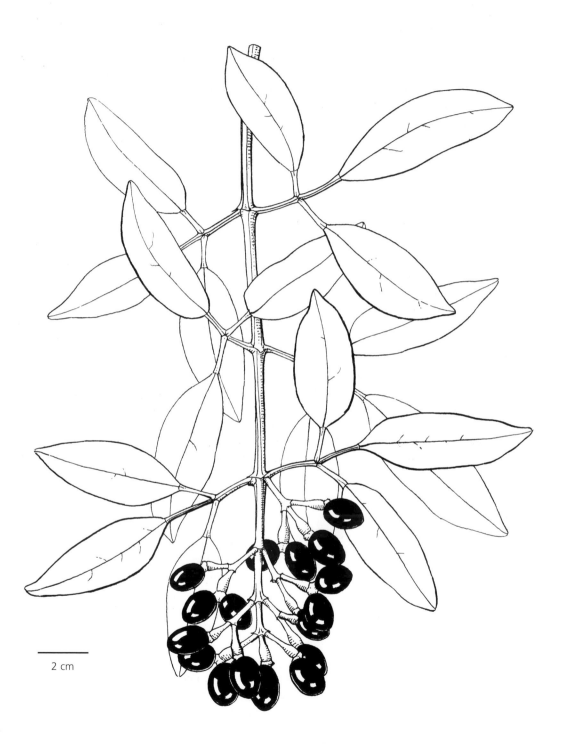

2 cm

Fig. 1. ***Jasminum lanceolaria***. Line drawing by Geoffrey Herklots.

Recent studies using DNA sequences have suggested a slightly different arrangement of the family. Wallander & Albert (2000) studied 76 species, representing all 25 genera of the family, using a cladistic analysis of DNA sequences from two non-coding chloroplast loci, the *rps16* intron and the *trnL-F* region. Non-molecular data, (chromosome numbers, fruit and wood anatomy, leaf glycosides, and iridoids), were also studied. Two genera, previously assigned to Verbenaceae or Nyctanthaceae, *Dimetra* (from Thailand) and *Nyctanthes*, (*Bot. Mag.* t. 4900), were shown to belong to Oleaceae, and be close to *Myxopyrum*.

Wallander & Albert (l.c.) suggested dividing the former subfamily Jasminoideae into four tribes: Myxopyreae (*Myxopyrum, Nyctanthes,* and *Dimetra*), Fontanesieae (*Fontanesia*), Forsythieae (*Abeliophyllum* and *Forsythia*), and Jasmineae (*Jasminum* and *Menodora*). They transformed the former subfamily Oleoideae into the tribe Oleeae, with subtribes Ligustrinae (*Syringa* and *Ligustrum*), Schreberinae (*Schrebera* and *Comoranthus*), Fraxininae (*Fraxinus*), and Oleinae (12 drupaceous genera, including the extinct *Hesperelaea*, from the Channel Islands in NW Mexico).

The genus *Jasminum* has over 200 species, mainly from Asia, Africa, Europe, the Pacific Islands and Australia; it includes tropical and hardy, deciduous and evergreen shrubs and climbing plants, with opposite, or sometimes alternate, simple, trifoliolate or pinnate leaves. The inflorescence is terminal, axillary or terminal on short lateral branches, in loose or dense cymes or flowers solitary. The calyx is 5-lobed and often very short. The corolla is usually scented, white or yellow, rarely red or pink, with a narrow corolla tube opening to 5, sometimes up to 9 or more, spreading lobes. The two stamens with short filaments are attached to the corolla and included within the tube. The ovary is 2-celled, each cell with 2 ovules. The fruits are fleshy, black or very dark coloured when ripe, and usually paired, but in most cases only one develops while the other aborts.

There is evidence that many species are heterostylous in a similar way to *Primula*, with thrum and pin-eyed plants. On different plants of the same species, the filaments of the stamens are attached at different heights within the corolla tube and the styles are of different lengths. This is a means of discouraging self-pollination and is associated with different lengths of corolla tube (and hence the appearance of the flower) and also with the size of the pollen grains.

Work carried out on *Jasminum fruticans* L. (Thompson & Dommée, 2000) and on herbarium specimens of other species, does show a variation in the length of the style and the position of the insertion of the filaments on the corolla tube. Many species in cultivation are a single clone, collected and introduced from the wild many years ago and vegetatively propagated ever since. This means that all plants derived from a single collection will have stamens inserted at a similar height. Many jasmines appear not to set fertile seed regularly, at least in northern European gardens, and this might be explained by the heterostylous nature of the flowers and their self-incompatibility. As plants from different collections become more widely distributed in cultivation and are grown together, it is possible that fruiting will become more frequent. Further detailed investigations would be required to ascertain if the heterostylous condition is typical of the genus as a whole.

The genus is mainly tropical and only a handful of species are reliably hardy garden plants in Britain and other areas with frosty winters; a few that come from the Sino-Himalayan region and one that occurs throughout the Mediterranean area, *Jasminum fruticans*, are hardy enough to survive the winters in warm temperate regions without winter protection.

There is as yet no satisfactory classification of the genus as a whole. Within it, one can identify three relatively small natural groups which may be split off and recognised as sections or perhaps even as subgenera. Two of these are unique in bearing yellow flowers: Sect. *Alternifolia* and Sect. *Primulina*.

The remainder, with white flowers, include the vast majority of species, comprising sections *Jasminum*, *Trifoliolata* and *Unifoliolata*, this last including more species than all the others put together.

The following paragraphs give an outline of the species, section by section.

SECTION JASMINUM (formerly called section Pinnatifolia)

This section, with its opposite and pinnate leaves, consists of five species and one interspecific hybrid. Amongst these is the original, type species of the genus, the so-called common or cottage jasmine, *Jasminum officinale*, (t. 4). Long cultivated and admired, it has often been confused with the very similar but more tropical *J. grandiflorum*, (sometimes seen under the name 'de Grasse'), which has slightly larger flowers borne in a cyme rather than subumbellate, and is grown under glass in temperate regions. Due to this confusion, the name *J. grandiflorum* has sometimes been wrongly given to a good form of *J. officinale*, the correct name for which is *J. officinale* f. *affine*.

As a native plant, *Jasminum officinale* has a distribution from the Caucasus, east through northern Iran to Pakistan and along the Himalayan range to south-eastern China. In contrast, *J. grandiflorum* is confined to northern Arabia (Map 1). Because of its sweetly fragrant flowers, it was taken early into cultivation, and there seems little doubt that it must have been carried by the Moors to southern Spain during their occupation of that area from the sixth to the fifteenth centuries (Green, 1995). *Jasminum grandiflorum* can sometimes be bought in garden centres in summer, as a group of cuttings around 10 cm tall, rooted and flowering in a pot.

The hardier *Jasminum officinale* has been grown in gardens in Britain for centuries. In 1548 William Turner in *The names of Herbes* said that 'it groweth communly in gardines bout London'. Within this species, as well as the typical green-leaved plant, there are two cultivars with variegated leaves: the robust 'Aureovariegatum' (or var. *aureovariegatum*) with golden splashed leaves, and the more delicate, white-variegated 'Argenteovariegatum' (or var. *argenteovariegatum*). Selected cultivars have been named 'Clotted Cream' (or 'Devon Cream'), 'Crûg's Collection' and FIONA SUNRISE ('Frojas').

Although most garden plants of *Jasminum officinale* are very uniform and rarely set seed (they possibly all belong to a single self-sterile clone) there is one with slightly smaller but equally attractive flowers, which is self-fertile and bears abundant, black, juicy fruits. This is cv. 'Inverleith', (t. 4), so named because the plant depicted in the portrait had been received from the Royal Botanic Garden in Edinburgh, Inverleith being the district of the city in which the garden is situated. Although no direct trace has been found, it was later discovered that an old plant that exactly matches this clone may be found in the National Botanic Garden at Glasnevin, in Dublin. It had been received there some 100 or more years ago, possibly via a French nurseryman, who had received it from one of the French missionaries then active in south-western China. 'Inverleith' is a desirable and floriferous plant, quite hardy in Britain, and propagates easily from seed.

Not as frost-resistant as *Jasminum officinale* is the Chinese spring-flowering *J. polyanthum* (t. 5 and cover), a species of borderline hardiness in Britain, but which in southern England will thrive against a south-facing wall, or elsewhere under glass. However many people will know it as a pot plant, commonly sold in florists shops or garden centres in winter and early spring, popular because

Plate 1. *Nyctanthes arbor-tristis.* Hand-coloured lithograph by W. H. Fitch from *Curtis's Botanical Magazine* t.4900 (1856).

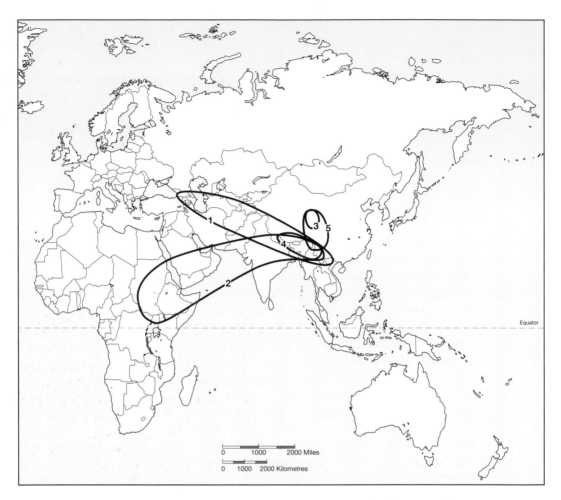

Map 1. Diagrammatic map of the distribution of section **Jasminum**: **1.** *Jasminum officinale*; **2.** *J. grandiflorum*; **3.** *J. polyanthum*; **4.** *J. dispermum*; **5.** *J. beesianum*.

of its sweetly-scented, attractive white flowers, pinkish in bud and on the exterior of the corollas. In Denmark, Holland and North America, the technique has been mastered to root young flowering shoots, or even inflorescences of this jasmine, and, as they come into flower, market them as house plants. Of the hundreds of such plants that are sold each year, one may be sure that very few will survive after the flowers fade; in fact inspection shows that some of them have no viable growing point — the whole of the growth having been taken up by the inflorescence. However, if they are grown in the open in a mild climate, they produce flowers which under cooler conditions are even more attractive, for the buds and exterior of the corollas are then a rich vinaceous red.

Another attractive pinnate-leaved jasmine is *Jasminum dispermum* (t. 6). It is little known in cultivation and scarcely hardy, so in Britain is best grown under glass. In the wild it is native to a wide sweep of the Himalaya, eastward from Pakistan, India to Tibet and western Yunnan (Map 1).

Jasminum beesianum (t. 7), is noteworthy as the only member of the genus with fully red flowers. Unfortunately the red is rather a dull one and does not stand out conspicuously, but the glossy black fruits that follow, usually in abundance (it is presumably a self-fertile species, in contrast to most others), mark the plant out as one worth growing. Although it has simple, undivided leaves it is undoubtedly closely related to *J. officinale*, with which it forms the hybrid, *J. ×stephanense*. This hybrid was raised in France at St Etienne (the French form of St Steven) and first marketed by Lemoine & Son of Nantes about 1920. It is not as robust as either of the two parental species, and is intermediate in character, producing flowers that are pink throughout, instead of red or white. Although this hybrid was raised in France, it was subsequently discovered growing wild in SW China and Tibet where both parental species are native.

SECTION TRIFOLIOLATA

This section is treated here in a traditional sense, that is to say, the species have been brought together because they possess trifoliolate leaves. Defined simply in this way it constitutes an easily recognised assemblage, a convenient, but artificial, bringing together of diversely related species — many of which have affinities with species that have simple leaves and are therefore outside this section.

However, in contrast to this traditionally simple but artificial assemblage of trifoliolate species, it should be noted that, although still containing the majority of trifoliolate species, the section has recently been delimited and redefined, based on a small group of five species centred in the Pacific and Australasia (Green, 2000). This small but natural group, typified by *Jasminum didymum* G. Forst., which was illustrated in *Curtis's Botanical Magazine* in 1878 (t. 6349), is distinct with its small corollas, the tubes of which vary between 2 mm and 12 mm in length, with lobes 2.5–7 mm long.

Although first described from Tahiti, *Jasminum didymum* has a wide distribution from Java and the Philippines eastwards to New Caledonia, Tahiti and the Austral Islands, with two subspecies in Australia. Being a tropical species it has to be grown under glass in Britain; though individually the flowers may be small, but they are produced in large paniculate inflorescences.

Four other species which have appeared in *Curtis's Botanical Magazine*, come into section Trifoliolata in the traditional sense. The first of these to be figured was *Jasminum azoricum* (t. 10) which, like *J. odoratissimum*, is a relic of the Tertiary period when the affinities of the floras of the Azores and Canary islands with those of tropical Africa were stronger than they are today. Although the epithet *azoricum* was given to this plant, there is no evidence that it ever grew wild in the Azores (an example of how, once a botanical name has been validly given, with priority, to a plant, it remains its correct name even though it may later be found to be misleading and inappropriate). Certainly, it is recorded from Madeira, although now extremely rare there, and very localised in the wild. Although frost tender, it has been grown in England under glass since the 18th century, and, because its sweetly-scented flowers are produced over an extended period, it is still met with occasionally as a greenhouse plant (sometimes, however, wrongly grown under the name *J. fluminense*, another quite different species).

A sectional classification of the genus *Jasminum* (based on whether the leaves are simple, trifoliate or pinnate) is artificial, as is clearly illustrated by *Jasminum angulare* (t. 8). Its leaves are nearly always trifoliate, but sometimes they may be simple, and occasionally even pinnate. *Jasminum angulare* is a widespread native of South Africa, variable in the degree of pubescence it may exhibit and notable for the length of its corolla tube, 2–3.5cm long. A selected clone has been given the cultivar name 'Anne Shelton', and is illustrated in the catalogue of The Plantsman Nursery (2005).

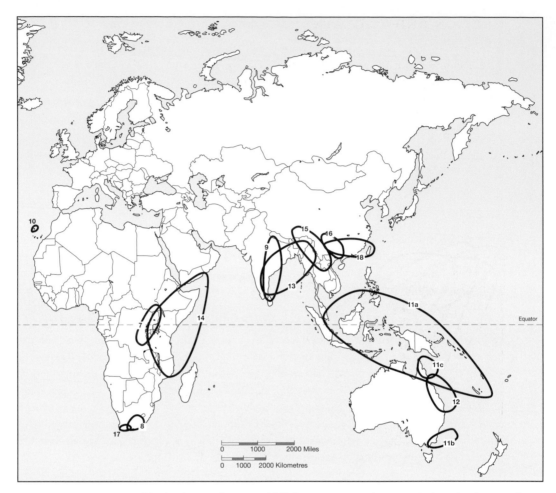

Map 2. Diagrammatic map of the distribution of section **Trifoliolata**: **7.** *Jasminum abyssinicum*; **8.** *J. angulare*; **9.** *J. auriculatum*; **10.** *J. azoricum*; **11.** *J. didymum*; **11a.** subsp *didymum*; **11b.** subsp. *lineare*; **11c.** subsp. *racemosum*; **12.** *J. dallachii*; **13.** *J. flexile*; **14.** *J. fluminense*; **15.** *J. lanceolaria*; **16.** *J. sinense*; **17.** *J. tortuosum*; **18.** *J. urophyllum*.

Jasminum urophyllum has a wide distribution in southern China, and might therefore be expected to exhibit considerable variation. Some forms of it have been given formal description as varieties, as for example var. *wilsonii* (t. 14). As a garden plant it is not hardy in temperate regions liable to frost, and although illustrated in *Curtis's Botanical Magazine*, it is but little known in cultivation, and hardly superior to other jasmines that have been illustrated.

Another trifoliolate species is *Jasminum sinense,* native over the whole of southern China (including Taiwan) (t. 12). Although it was illustrated in *Curtis's Botanical Magazine* as recently as 1993, it had unknowingly been grown in the Palm House at Kew for over a hundred years (Green, 1993). Nevertheless, it can hardly be said to be well-known in cultivation. With densely tomentose foliage, a corolla tube 3–4 mm long, and sweetly scented flowers, it can, however, be recommended for cultivation in warm temperate or tropical gardens.

SECTION UNIFOLIOLATA

Probably no other jasmine has been cultivated for so long, so widely, and in such quantity, as *Jasminum sambac*, a name derived from the ancient Persian name *zambac*. It is a frost tender species, almost certainly a native of the warmer areas of India and has been in cultivation there for centuries. To judge by the literature, numerous cultivated varieties exist with native Indian names. These are not known in the west, however, where two forms are commonly cultivated: one, known as 'Maid of Orléans', has a simple double corolla and the other, 'Grand Duke of Tuscany' (t. 21), has the

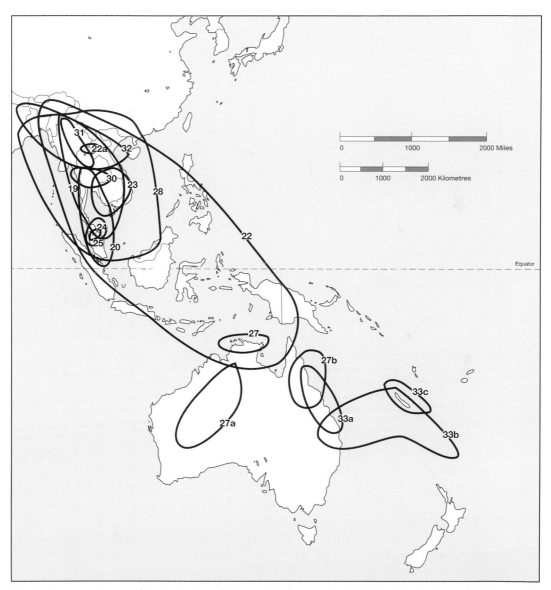

Map 3. Diagrammatic map of the distribution of Asian and Australian section **Unifoliolata**: **19.** *Jasminum adenophyllum*; **20.** *J. decussatum*; **22.** *J. elongatum*; **22a.** *J. perissanthum*; **23.** *J. harmandianum*; **24.** *J. kedahense*; **25.** *J. maingayi*; **27.** *J. molle*; **27a.** *J. calcareum*; **27b.** *J. kajewskii*; **28.** *J. multiflorum*; **30.** *J. nobile*; **31.** *J. scandens*; **32.** *J. syringifolium*; **33.** *J. simplicifolium*; **33a.** subsp. *suavissimum*; **33b.** subsp. *australiense*; **33c.** subsp. *leratii*.

corolla lobes are greatly multiplied. Having been developed and cultivated for so long, these plants often produce atypical leaves which may be alternate or ternate (as in the plate, where the plant was given the name var. *trifoliatum*).

Because of their particularly sweet fragrance the flowers of *Jasminum sambac* (and sometimes the following species too) are commonly strung together to make a much prized garland or *lei*, traditional in India but also popular elsewhere, for example in Hawaii. *Jasminum sambac* must have been an early introduction to China, where today it is extensively cultivated for its dried flowers, which when added to ordinary China tea, produce fragrant jasmine tea called *muli cha*, now sold and widely enjoyed in the West. *Jasminum sambac* was also grown in the south of France where the flowers' sweet fragrance was extracted by a process known as *enfleurage*. In this, newly opened buds are gathered early in the morning and spread out on trays of refined lard; this captures the essence, and from it the jasmine scent is then extracted by distillation. This process is described in more detail in the chapter on jasmin in the perfume industry (see p. 23).

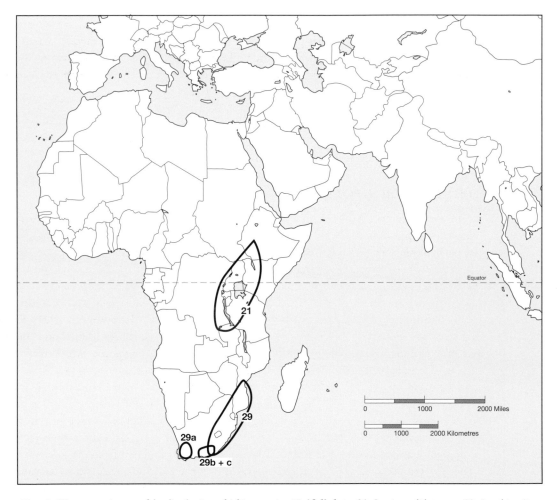

Map 4. Diagrammatic map of the distribution of African section **Unifoliolata**: **21.** *Jasminum dichotomum*; **29.** *J. multipartitum*; **29a**. *J. glaucum*; **29b.** *J. streptopus* var. *transvaalense*; **29c.** *J. stenolobum*.

Often closely associated, and likewise tropical and frost tender, is the equally well-scented *Jasminum multiflorum,* which has been cultivated in India and adjacent countries since ancient times. It appeared in *Curtis's Botanical Magazine* twice, once in 1818 under the synonym *J. hirsutum* (t. 1991), and again in 1881 under the name *J. gracillimum* (t. 6559); they represent two of the forms of this species in cultivation. The dried corollas of *J. multiflorum,* like those of *J. sambac,* are also used to flavour tea in China.

Jasminum simplicifolium (t. 22) was an early introduction from Australia, before 1806. It is not as floriferous as the previously mentioned species, and today is found in few collections. Likewise, *Jasminum maingayi* is a tropical species, this time from Penang, Malaysia, and hardly, if at all, known in cultivation today. The same may be said of *Jasminum kedahense* (t. 16), which is also a native of Malaysia.

Few jasmines have such large flowers as *Jasminum nobile* subsp. *rex* (t. 20). These flowers measure 5–7 mm across, and are strikingly pure white; their impact is enhanced by the dark green foliage. Unfortunately for gardeners in temperate regions, this is a tropical plant from Thailand, and in Britain at least, has to be grown under glass. Also unfortunate is the fact that the flowers have no scent, presumably because they have no need for this as their size and gleaming whiteness are already sufficient to attract potential pollinators. The challenge here, for someone with access to living plants of this jasmine, and perhaps others like the fragrant *J. multiflorum,* was to introduce scent through hybridisation, and this has recently been done by Paul Swedroe of Fort Lauderdale, Florida, who has produced the hybrid 'Ann Clements' (see p. 110).

Jasminum nobile subsp. *nobile* is native in Burma (Myanmar), but although first described in 1882, it is still little known. Further collections from the area may produce plants which bridge the differences between this and subsp. *rex,* as dried specimens from Cambodia, Laos and Vietnam already indicate that we are really dealing with a single species (Maps 3 and 4).

SECTION ALTERNIFOLIA

This section may be recognised by its combination of yellow flowers and alternately arranged leaves, the latter an exception for the whole of the family Oleaceae, to which *Jasminum* belongs.

Jasminum odoratissimum was the first member of this section to be included in *Curtis's Botanical Magazine* (t. 285), in 1794. The name is unfortunate, for it is far from being the most fragrant species. Though native of Madeira and the Canaries, its affinities lie in the tropics and not with the Mediterranean, as is the case with many other Macronesian endemics (see also *J. azoricum*). In particular, although geographically widely separated, its closest relationship is with *Jasminum odoratissimum* subsp. *goetzeanum,* a native of eastern Africa (Tanzania, Kenya, Zaire and Zambia).

The only jasmine native to Europe is the Mediterranean *Jasminum fruticans* which is distributed from Portugal, in the west, to Turkey, Iran and the Crimea in the east, and along the North African coast. It forms an evergreen or semi-evergreen upright shrub to 2 m tall, and though its bright yellow flowers contrast well with the dark green of its foliage, it is not as showy as most jasmines and the flowers generally lack scent. Understandably therefore, it is not often met with in gardens.

In 1815, when describing *Jasminum revolutum,* another member of this section, Sims thought he was dealing with a new species, but his plant was not specifically different from the *J. humile,* a name established by Linnaeus in 1753. With a wide distribution from Afghanistan and Pakistan, in the west, to China (Yunnan and Sichuan) in the east, *J. humile* is a variable species that encompasses *J. revolutum* and the little known *J. reevesii.* Nevertheless, the jasmine typified by Sim's plate is somewhat distinct from this variable species, and with its rather larger flowers, is worthy of recognition as a cultivar, 'Revolutum' (*Curtis's Botanical Magazine* t. 173). It is not as hardy as some other introductions of *J. humile* and, in most of Britain, is best grown under glass.

Jasminum humile has been introduced to cultivation several times and there is now considerable variation amongst plants in gardens. A smaller leaved form with more divisions to the leaf is f. *wallichianum,* while another aspect of *J. humile,* which is occasionally cultivated, was named and described as *J. farreri* by John Gilmour in *Curtis's Botanical Magazine* (t. 9351), named in honour of that outstanding plantsman, Reginald Farrer (1880–1920). Its main distinctions rest in the pubescence of its young branches, leaves and inflorescence. If and when it requires a distinct name it may be referred to as forma *farreri.*

Exciting because of its unexpected discovery as recently as 1995, is one of the newest species of jasmine to have been described, the remarkably distinct *Jasminum leptophyllum.* It was discovered in the Palas Valley of Kohistan by the Pakistani botanist Rubina Rafiq, who described and named it

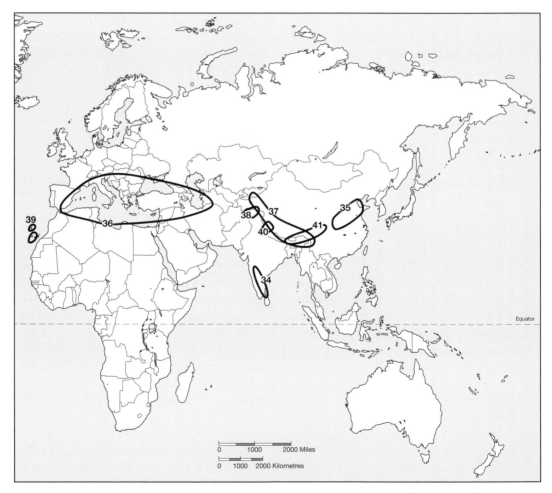

Map 5. Diagrammatic map of the distribution of section **Alternifolia**: **34.** *Jasminum bignoniaceum*; **35.** *J. floridum*; **36.** *J. fruticans*; **37.** *J. humile*; **38.** *J. leptophyllum*; **39.** *J. adoratissimum*; **40.** *J. parkeri*; **41.** *J. subhumile.*

the following year. Alongside its large yellow flowers, perhaps the most remarkable character is its possession of simple, linear or linear-lanceolate leaves, quite distinct from those of any other member of this section. Fortunately, viable seed, which germinated, was obtained by Mike Sinnott and others from Kew who were in the Palas Valley in Northern Pakistan at the time Dr Rafiq was carrying out her field work there (Sinnott, Rafina & Green, 2000).

Another Himalayan species of limited distribution also belongs to Sect. *Alternifolia*. This is *Jasminum parkeri,* which is neither a climber nor a scrambler but forms a compact, almost cushion-shaped, plant, ideal for cultivation in a rock garden. As a wild plant it has a limited distribution in Chamba State in Himachal Pradesh.

Jasminum floridum and *J. bignoniaceum* should also be included amongst the alternate-leaved jasmines in cultivation. The former is a native of central China, and the latter endemic to the Nilgiri and Palni Hills of southern India. *Jasminum floridum* was described in *Curtis's Botanical Magazine* as long ago as 1883. In cultivation it is usually grown against a wall, but to judge from its native area of distribution in north-central China it should be quite hardy in Britain. *Jasminum bignoniaceum* has, surprisingly, survived in the author's own suburban garden in southern England against a south-facing fence, and has persisted through many winters against a warm wall at Kew.

Finally, within this section, mention should be made of *Jasminum subhumile*. It is recorded as native from SW China (Yunnan and Tibet), India (Assam and Manipur) and Myanmar, and two or three varieties of doubtful significance have been described (Map 5).

SECTION PRIMULINA

This small section contains only two species, and is easily recognised by its yellow flowers and opposite leaves (in contrast to the yellow flowered species in Sect. *Alternifolia* where, as the name indicates, the leaves are borne alternately on the stems).

The better of the two species is the popular Winter Jasmine, *Jasminum nudiflorum* (t. 31), which was first introduced from China, its native country, by Robert Fortune in 1844 (Map 6). Probably no other plant that flowers in mid-winter enlivens our gardens as much as this jasmine. Under a continental climate and during a really cold period the flower buds remain dormant until the beginning of spring when they come rushing out together, but, in Britain, given a warm spell, they may open at any time in January and February — individual flowers may even open intermittently from November onwards.

In addition to providing unseasonable colour for our gardens, *Jasminum nudiflorum* will grow in almost any soil, whatever its exposure, and can be propagated very easily by layering. In fact most, if not all, jasmine plants seen in gardens have been propagated in this way. If, as with most jasmines, *J. nudiflorum* is self-incompatible, this might explain why seed is never set — the plants one sees in suburban gardens, may for instance belong to one immense clone, one vast genetic 'individual', scattered in thousands of gardens. Most of the plants of this species that have had their chromosomes counted have turned out to be tetraploid, although in China some aneuroploid numbers have been recorded as well.

Closely related to winter jasmine is primrose jasmine, the other member of this small section, *J. mesnyi* (t. 30) or *J. primulinum* as it was once and appropriately called (although the former name has nomenclatural priority and is therefore the correct one). The latter name, alluding to the primrose, is quite descriptive of the plant's flowers, while the former was given to it in honour of William Mesny (1843–1919), who became a General in the Chinese Imperial Army and who, during his travels, occasionally collected plants, including this jasmine. It is nowhere near as hardy as *J. nudiflorum*, but grown against a south-facing wall, will survive and flower in southern England. Even in such a position, however, the flowers which appear in the spring are liable to be spoilt by frost. To see it at its best it should be grown in a cool greenhouse or conservatory when, in March or April, it will cover itself with large cheerful semi-double blooms. Alternatively, it is sometimes, and very effectively, grown as a standard, 4–6 feet high, with the flowering shoots hanging down all around and providing a circle of blossom. It too seems not to produce seed, and perhaps, once again, this can be attributed to the clonal propagation of a genetically self-sterile plant.

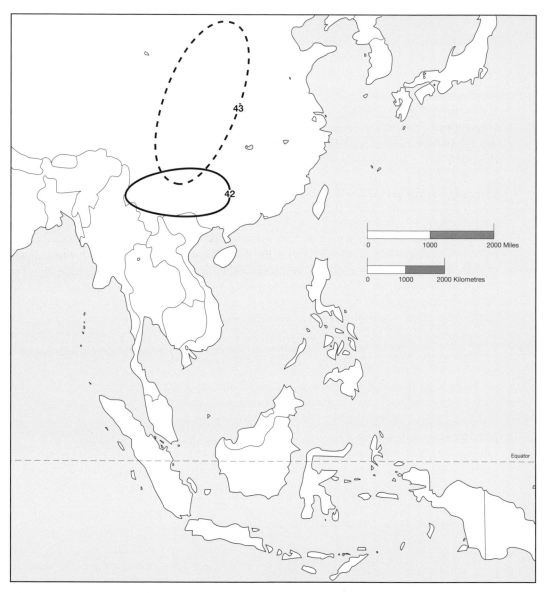

Map 6. Diagrammatic map of the distribution of section **Primulina**: **42.** *Jasminum mesnyi*; **43.** *J. nudiflorum*.

2. JASMINE IN CULTURE

HISTORY

Many jasmines were highly valued and cultivated in ancient times for their scented flowers, and apart from their ornamental value, were grown commercially for the extraction of scent, or for flavouring. Initially jasmines were used in their native lands, but several species moved around the world along ancient trade routes. It is assumed that *Jasminum sambac* reached China from India and *Jasminum grandiflorum* was taken along the old Silk Road eastwards from Iran; they are both mentioned in a treatise by Hsi Han (263–307), the Nan Fang Tshao Chu Ang which dates from 304, as flowers transplanted from the western countries by foreigners and grown in Kuangtung. In Chinese there are names for two kinds of jasmine, yeh-hsi-ming, derived from the Persian, for *J. grandiflorum* and moo-li, derived from the Sanskrit mallika, for *J. sambac*. Both names are thought to date back to the second century BC (Needham, 1986).

GREEK AND ROMAN USE OF JASMINE

Jasmine was used as a perfume by the Romans, but it is probable that they imported it from the Middle East, as neither the Greeks nor the Romans seem to have known how to extract and fix it. In Greek, jasmine was recorded first by Pedianos Dioscorides in c.100 AD when describing a Persian perfume which he called ίασμη, ίασμελαιον or ίασμινον μύρον. Melaion or eΐlaion was oil, particularly olive oil, whereas μύρον, was any oil made by distillation, and generally applied to perfume. Dioscorides (Aët.1.119) states that this was prepared "eΐk ton aΐnθn ton leukon tou iΐou, kai eΐlaiou sesaminou" (from the white flower of the violet and oil of sesame). It is not clear whether there is confusion here between the white violet and the jasmine. Both were used to make scent, and the flowers of the two, when picked, are rather similar. Dioscorides was a military physician who travelled with Nero's army, and wrote a medicinal herbal which was widely regarded from Hellenistic times onwards and unsurpassed until the Renaissance.

The Latin *jasminum* is not found in classical authors, which is surprising as it is likely to have been imported to Rome from Egypt or the eastern Mediterranean, as were large quantities of roses. Pliny records the story that the Roman orator, Lucius Plotius Gallus, childhood friend of Cicero and later of Mark Antony, while hiding from his political enemies, was betrayed by the strong scent of which he was inordinately fond. This was during the second Triumvirate in around 43 AD, and it has been suggested that the scent was imported jasmin.

PERSIAN AND ARABIC POETRY

Jasmines have also been the inspiration of poetry from the ancient Tamils over 2,000 years ago to Sanskrit, through Persian and Arabian poets in the Middle Ages, to Europe and the modern day. In Persian poetry, the word used for Jasmine is *siman*, which came from the Persian *ya⁻sam*, *yasamin* or *yasaman* (the current name in Farsi); this then became *yasmin* in Arabic, from which was derived

the French Jasmin. Rose, hyacinth, *Narcissus tazetta*, violet and jasmine are mentioned regularly, all scented plants which tended to be contrasted with those which were more colourful, but without scent (Thackston, 1996).

In *Divan 116, line 2,* the poet Hafiz contrasts jasmine with the Judas tree (*Cercis*):
'The Judas tree will give an agate goblet to the Jasmine;
The eye of the Narcissus will gaze at the anemone'.

Later, it is contrasted with the anemone:
'As balm for the anemone's scar,
the jasmine has freely given its scent'   Kalim, *Qasa'id Q. 29:14.*

Hafiz Shirazi (1319–1389) was a teacher, philosopher and poet, who lived in Shiraz and Isfahan. He was for a time court poet to Abu Ishak. Kalim (d. 1651) was the court poet of Shah Jahan, son of Babur, who conquered India in the early 16th century and founded the Mughal dynasty.

EUROPE

The Persian *Jasminum grandiflorum* was introduced into Europe by the Moors over 1,000 years ago, having been taken westwards through North Africa to Spain. There is a good illustration of *Jasminum officinale* in a herbal, now in the British Library, which was written in Salerno between 1280 and 1310 (Egerton MS 747, folio 98); it is labelled Sanbaco or Gessominu (Fig. 2). The text of Egerton 747 which accompanies the drawing is *Tractatus de Herbis*, which was a compilation from other herbals including Dioscorides. This illustration, and others in Egerton 747, are early examples of more life-like representations of plants and appear to have been drawn from nature or copied from a field sketch, rather than copied from generations of earlier herbals (Collins, 2000).

Sir Thomas Hanmer, writing in 1659, knew *Jasminum grandiflorum* as the Spanish or Catalonian jasmine and regarded it as tender in England. Its cultivation as a crop in Europe began in the seventeenth century in south-eastern France around Grasse, still the centre of the perfume industry (see page 23). The Mediterranean *J. fruticans* is illustrated in Alexander Marshall's florilegium which dates from around 1660, but it is recorded in gardens before the end of the sixteenth century.

FOLKLORE

There is folklore attached to jasmine flowers in several countries and they are frequently associated with wedding ceremonies. Jasmine is also very significant in the Language of Flowers, which became so important in Victorian England, although the meanings may have originated in India. Different species had different significance. Yellow jasmine flowers indicate grace and elegance, whereas others denote amiability, sensuality or attachment.

Certain species have been designated National Flowers. The Philippines adopted *Jasminum sambac* in the 1930s where it is known as sampaguita, and the Indonesian government adopted it as melati, in 1990. It is said to symbolise purity, eternal love and nobility. Pakistan chose the common species *J.*

Plate 2. *Jasminum grandiflorum*, hand-coloured engraving from *Botanical Register* t.91 (1816). Artist unknown.

Fig. 2. **_Jasminum officinale_** L. with _Thalictrum flavum_ L. (left) from Egerton MS 747 (British Library).

officinale, known locally as chambeli. An even more exotic use for an African species, *J. angulare,* is as protection against lightning by certain peoples. More practically the stems of jasmines have been used for wicker work and bindings in Laos and made into rope or used as ties in Kenya. Juice extracted from the fruits of *J. humile* provides a black ink and the juice of *J. fluminense* is used for colouring baskets.

Of course, it is the scent of the flowers which makes them so highly valued in religious ceremonies, in food flavouring, in pot-pourri, flower waters, cosmetics, soaps and toiletries. It is possible to create a simple perfume by adding flowers to oils such as sesame, and this is done on a large scale in aromatherapy and in the perfume industry (see page 28). In several countries in south-east Asia the flowers are used for personal adornment, especially in the hair, or worn as garlands.

RELIGIOUS IMPORTANCE

It is believed that jasmine flowers were used in offerings to the gods in Ancient Egypt. Later, in the fourth century AD when Patna, in Sri Lanka, converted to Buddhism, there are references to the abundant planting of jasmine with lotus in courtyards (Goody, 1993) and still today garlands or posies of flowers are offered in temples or even used as a talisman in cars in Buddhist culture. In Malaya, the scent is said to have a particular attraction to the spirits and is used to summon their help or in the exorcism of disease.

In Hinduism, it is thought that offerings of flowers replaced the former blood sacrifices. *Jasminum sambac* flowers are sacred to the Lord Vishnu and are used as offerings in Hindu ceremonies. The flowers are considered to be one of the darts of Kama Deva, the god of love of Hindus (Pandy, 1989) and are widely used in garlands. They are especially important in wedding ceremonies in many countries where the plant grows, including India, Indonesia and Pakistan where the bride and groom as well as their surroundings are richly decorated with jasmine. As incense, jasmine-scented oil is mixed with a carrier such as powdered wood.

A ritual in China was to give flowers as gifts for the New Year, but eighteenth-century Beijing was too cold to grow flowers and it is reported that jasmine was forced in kilns. In *Rixia Jiuwen,* Zhu Yizun, who lived in the seventeenth century, describes how jasmine and other plants such as peonies and peach blossom were forced for the New Year by planting them in pots in a cavity below ground and heated by burning horse dung. This was method used from the Han dynasty 202 BC–220 AD (Goody, 1993).

GARLANDS

The tradition of making garlands as forms of decoration, of greeting or for expressing reverence is timeless. In Hawaii this tradition has grown into a big industry. The flowers bear the name 'Pikaki' (meaning white peacocks), given by Princess Kailuylani of the now extinct royal family of Hawaii, who was fond of both it and the white peacock.

Jasminum sambac is especially important as a wedding flower lei, or garland, and harvested all year round, although less in December to February. The flowers are picked in early morning and refrigerated in sealed glass jars until sufficient have been collected. A special needle is used to thread the blossoms which can be made into a simple garland which requires 100 blossoms for a 40 inch lei. Over 600 blossoms may be used in the preparation of a double and more complicated rope pattern.

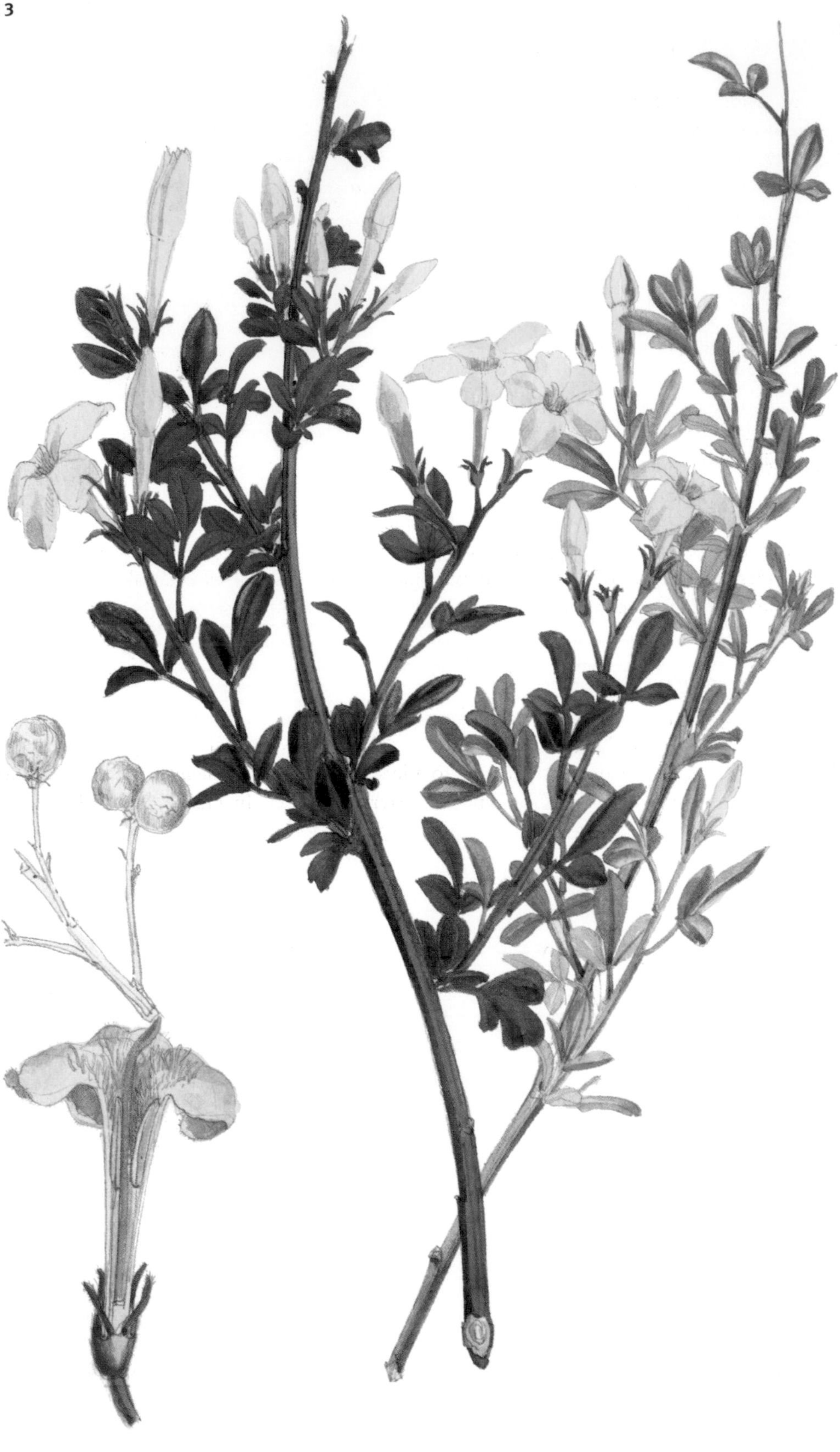

In China garlands have been known since the seventeenth century. In India the buds of *Jasminum sambac* are picked in late afternoon and open 7–9 hours later when they are sold and the garlands made up in the flower markets for decoration or hair adornment. Piles of loose flowers rather than sprays or branches may be seen in the markets. However, in Malaysia the flowers are harvested as buds during the day and may last several days if kept cool.

FLAVOURING

There is a small but significant use of jasmine flavouring in the food industry, in alcoholic and soft drinks, confectionary and some desserts. They have also been used to flavour tobacco. A syrup can be made by layering the flowers with sugar which is again used for flavouring or diluted as a drink. Their main use for flavouring is in jasmine tea which is especially popular in China where it is believed to have been drunk since the Sung Dynasty over 700 years ago. The two species used for this purpose are *Jasminum sambac* and *J. grandiflorum* and China is the main producer. To flavour the tea, the flowers are layered with green tea until the tea has absorbed the scent. Slightly different processes are used for different grades and hot air may be used to dry the leaves more quickly. The best grades are thought to be made of flowers picked during May. The spent flowers will no longer retain any scent so are either removed or used as decoration.

FLOWER MARKETS

The commercial growing of jasmines is limited to a very small number of species, the main ones being *Jasminum grandiflorum*, *J. sambac* and, to a lesser extent, *J. auriculatum*. One or two others may be grown on a very small scale in some far eastern countries. In France, Spain, Italy, Morocco, Algeria and especially Egypt, *J. grandiflorum* is grown for the flowers which are used for the production of the essential oil. *J. sambac* are grown for oil production and there is a now a large export trade in fresh flowers from Bombay to the Middle East and Gulf States, totalling 30,000 kg in 1995 (Weiss, 1997).

Plate 3. *Jasminum fruticans* L. Hand-coloured engraving by Sydenham Edwards for *Curtis's Botanical Magazine* t.461 (1799).

3. JASMIN IN THE PERFUME INDUSTRY
— AN OVERVIEW by Robert Calkin

Of the natural floral products used in perfumery, jasmin (the French spelling is commonly used throughout the fragrance industry) and rose are the two most important. Among others which are also widely used are tuberose, orange blossom, narcissus, mimosa and *Osmanthus*. Although synthetically produced aromatic chemicals now make up the bulk of perfumery formulations, the natural products, in spite of their high cost, still find a use in fine perfumery.

Jasmine oil absolute is usually a dark orangey-brown viscous liquid, sometimes reddish, and darkening in sunlight, although it differs slightly depending on its source. The scent is very floral and long lasting. The oil contains about 300 different components compared with about 400 in rose oil, some in only minute amounts and it is this complexity which gives it a unique perfume. The oils extracted from different jasmine species have slightly different compositions but the overall effect is subtle and only an expert 'nose' would be able to recognise them. However the proportion of these constituents changes as the flowers are picked and while awaiting processing, as well as according to the time of year when they are harvested. They are sufficiently different for each species and geographical origin for all these factors to be analysed chemically and distinguished.

The significant components include indole, (which is in a higher concentration in oil produced from *Jasminum sambac*, hence the heavier scent), benzyl acetate, *cis*-jasmone, methyl jasmonate, linalool, phenylacetic acid, benzyl alcohol, farnesol, jasmolactone and methyl anthranilate (Harborne & Baxter, 2001).

Of the two types of jasmin absolute produced from *Jasminum grandiflorum* — that produced by the traditional enfleurage process and that produced by solvent extraction, only the latter is today in general commerce. Small amounts of jasmin absolute 'chassis', produced by the enfleurage process may still be manufactured for use in long established perfumes (formulation secrecy makes it impossible to know in which), but its very high price probably rules it out for use in new formulations.

Most of the absolute produced by solvent extraction, which originally came from the region of Grasse in the south of France, now originates from Egypt. Much of the 'concrete' produced by the initial solvent extraction is still sent to Grasse for the production of the absolute – sometimes called, misleadingly, 'Jasmin Absolute de Grasse'. True Jasmin de Grasse is still regarded as the finest quality, though today, because of the high cost of production, its use is probably limited to a few perfume houses such as Chanel, who own the fields and control its production. It is a remarkable fact that almost any plant grown in the Grasse microclimate and soil, a few miles inland from Cannes, produces a finer fragrance than that grown elsewhere.

The cost of jasmin absolute comes largely from the labour involved in picking the flowers. This has to be carried out in the early hours of the morning, when the fragrance is at its best, and is frequently done by children, who are the right height for picking the flowers and less expensive to employ. There is some disquiet in the industry about the conditions under which such children are employed.

Until 50 years ago jasmin absolute was often a major constituent of fine perfumes, being used at up to 4% of the formulation, or even higher: *Joy* of Jean Patou and *No. 5* of Chanel are notable examples, but there are many others. There was probably no important perfume that did not contain it. It contributed both to the perceived fragrance and to the richness and sense of quality. Jean Carles, a famous perfumer and teacher working in Grasse in the 1940s and 1950s used to say that jasmin absolute was to perfumery what butter was to haute cuisine. Nothing gives quite the same effect. It also acts in the same way as butter in giving 'finish' and 'roundness' to the composition. (The vocabulary of perfumery is hopelessly inadequate!)

The composition of a fine fragrance demands a balance between simplicity in its overall structure and complexity in the number of ultimate components. It is this complexity which gives the impression of richness and quality. Jasmin absolute contains over 300 component chemicals, many in only trace amounts, and it is this complexity which is one reason for its value in perfumery. Even in very small amounts it can give a 'natural' character to an otherwise synthetic mixture.

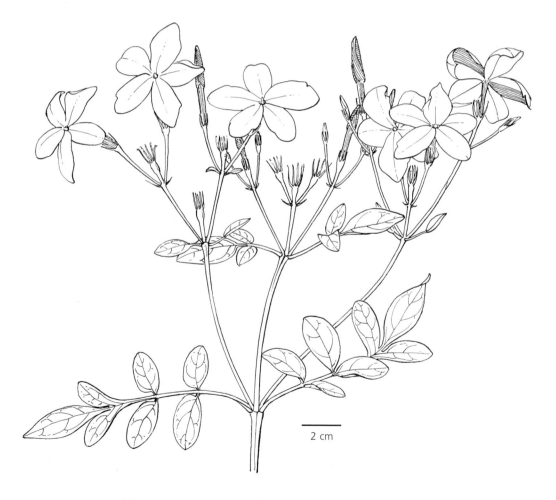

Fig. 3. ***Jasminum grandiflorum***. Line drawing by Geoffrey Herklots.

Similarly, rose oil contains some 400 components. (More than 500, however, have been identified in beer!) A remarkable attribute of these natural fragrances is the way in which this complexity produces such a unified identity (a rose smells precisely of a rose, a jasmine of jasmin). Co-evolution with pollinating insects seems to have selected a structure of fragrance which is appreciated also by our own species.

Although there are many products on the market, based on a chemical reconstruction of the natural product, none fully succeeds in capturing either the fragrance or richness of the original. So-called jasmin 'bases', which are largely synthetic and to some extent replicate the fragrance of the flower, are widely used by perfumers as building blocks in their perfumes. A small amount of the natural absolute is sometimes added to these 'bases', as well as to those replicating the fragrance of other flowers, to give them a touch of 'quality'. The biggest single ingredient in jasmin absolute is benzyl acetate, one of the cheapest to produce of all synthetic perfumery materials, and this is used as the starting point for the creation of nearly all jasmine-type creations.

The fragrance of jasmin is deep, warm, sensuous and erotic. It is the fragrance of the night. All successful perfumes contain an element of the animalic and it is this aspect of the fragrance of jasmin and its ability to blend with other animalic notes, such as musk and civet, which contributes to its value to the perfumer. An important constituent of the fragrance is indole (C_8H_7N), the chemical responsible when oxidised for the red staining of the flowers when damaged. At high concentrations it has a powerful faecal character, which most people find unpleasant; but at low levels it is perceived as sweetly floral. Indole occurs in animal faeces and its presence in many flowers, notably also in orange blossom and lily of the valley, no doubt acts as an attraction to insects. In jasmin absolute, although present at the relatively high level of 2.5% it is so well blended into the composition as to be acceptable to the human nose.

There is also a fruity character to the jasmin fragrance, coming from the presence of lactones, which gives it the ability to blend with other fruity notes such as peach, as in the famous perfume *Femme* de Rochas. No doubt many of the fruity perfumes which seem to dominate the market today contain small amounts of jasmin absolute. It is also an indispensable part of the classic 'chypre' type of formulation.

Another important ingredient of the fragrance is methyl jasmonate. This material was first identified in jasmin by the Swiss company of Firmenich in the 1960s and was synthesised and sold in its dihydro- form under the name of Hedione. No aroma chemical can be said to have had a greater impact on the perfumery industry. Smelled on its own the fragrance is perhaps disappointing with only a light, slightly fatty jasmin character. Indeed it took perfumers some years to recognise its potential. However it combines a wonderful transparency (again words are inadequate) and persistence with extraordinary diffusion. It has the ability to give 'life' to a perfume. Its first use was in *Eau Sauvage* by Christian Dior, launched in 1966. Today almost no fine perfume, or indeed any fragrance, including those for household products, is without it. It is used at up to 30% in the formulation of some deluxe perfumes. It can truly be said to have revolutionised perfumery. The presence of methyl jasmonate in the absolute, at about 23%, no doubt makes an important contribution to its overall fragrance and was one of the reasons for its traditional use.

The combination of indole and methyl jasmonate must also play a part in the amazing diffusiveness of the fragrance of the flower in nature. A field of jasmin can be smelt from many miles — a fact no doubt appreciated by insects. As with most white flowers pollination probably occurs mainly at night, by night-flying insects, when the fragrance is at its best.

Methyl jasmonate has some other remarkable properties in the plant world, where it occurs quite widely, as a chemical messenger and in triggering the plant's defence mechanisms. It is also reported as being used for prolonging the shelf life of fruit.

The traditional species used for the production of the absolute for perfumery is *Jasminum grandiflorum*. In recent years, however, the absolute from *J. sambac* has also come to be used. In the past this species, coming from China and India, was used mainly for the flavouring of jasmine tea. The fragrance is less fine than that of *J. grandiflorum*, being 'greener', due to a higher level of hexenols (which smell of cut grass and unripe banana), and more indolic. However it is considerably cheaper than the traditional product and is now used in much the same way to give a natural character to otherwise largely synthetic compositions.

Both types of jasmin are used in aromatherapy, where no doubt the presence of indol, with its particular associations, and methyl jasmonate work at a subjective level in giving the claimed effects.

Many of the 'noble' materials used in the perfume industry find their way also into synthetic flavours and as a boost to other natural products. Jasmin absolute is certainly used by flavourists for the same qualities that it brings to perfumery, though again, because of secrecy, it is impossible to give any specific examples. It would seem a natural for use in fruit flavours such as apricot and peach.

Fig. 4. Samples of Jasmin from different species of *Jasminum*. From left: *Jasmine* sp.; *J. pubescens* Willd.; *J. grandiflorum*; *J. sambac*. From the Kew Economic Botany collection.

HISTORY

The cultivation for oil from *Jasminum grandiflorum* was once centred around Grasse in the French Alpes Maritimes, with the main period starting around the seventeenth century and peaking between the two world wars. However, with the very high labour costs as well as competition for the land, cultivation for the industry moved to other countries bordering the Mediterranean including Algeria, Morocco, Italy and Spain. In 1912, Charles Garnier started experimenting with plantations near Cairo (Guenther, 1952) and now Egypt is the centre of production of the 'concrete' produced by solvent extraction, which is then returned to France for further processing. Plants in the old French nurseries were grafted onto *J. officinale* to give a hardier plant but once the industry moved to the warmer Egyptian climate, this was no longer necessary, thus reducing labour costs still further. However, true Jasmin de Grasse grown in France is still regarded as the finest quality. *Jasminum sambac* has been cultivated for its flowers but more recently it, together with *J. auriculatum*, has been used for the extraction of oil for perfumery.

COMMERCIAL CULTIVATION

Jasmines will grow in almost any well-drained soil as long as it is not waterlogged or saline, in sun and at a pH of between 6 and 8. They need regular feeding to increase flower yields using farmyard manure or chemical fertilisers, such as ammonium sulphate phosphate and potassium, and irrigation can lengthen flowering period (Weiss, 1997). They are climbing plants (except for some cultivars of *Jasminum sambac*) and need some sort of support similar to vines. They are pruned after flowering so as to maintain a height convenient for picking. Propagation is vegetative (by layering or cuttings) to maintain selected high-yielding cultivars and for faster establishment of the young plants. A new plant should be giving a good flower crop within 3 years and continue for 10–12 years. Although, as an ornamental, jasmines suffer from few diseases, as a monoculture they are susceptible to root borne disease.

Jasminum grandiflorum grows in the milder climate of Northern India as well as Egypt and other Mediterranean countries. *Jasminum auriculatum* and *J. sambac* are grown in more southerly and warmer climates as both require higher temperatures and rainfall; although selected strains, especially of the latter species, are chosen for their adaptation to different local climates and intended use (oil or flower production).

HARVESTING

The cost of jasmin absolute, the final purified essential oil, comes largely from the labour involved in hand-picking the flowers, which makes it one of the most expensive oils. Picking has to be carried out in the early hours of the morning when the fragrance is at its best. When the half-opened and fresh fully-opened flowers are picked they must not be bruised as doing so releases the indole, resulting in a different scent. Other chemical constituents also change once the flowers are picked so processing needs to be carried out nearby. Continual picking through the season ensures the continuation of flowering for as many months as possible.

It is estimated that a skilled worker can pick about 5,000 flowers in an hour. As 10,000 flowers give a kilogram of flowers and 8,000 flowers are needed to produce 1 gram (or 25 drops) of oil, the high price is not surprising.

The yield is a combination of the number of flowers produced per hectare and the amount of oil which can be obtained from them. But the resultant concrete is also affected by species, the cultivar, the season of picking, the interval between picking and processing, and the quality of flowers. *Jasminum auriculatum,* cultivated in India especially for flowers, can produce about 2,600 flowers per kilogram from April to December, producing about 2,000 kilograms per hectare. *Jasminum sambac*, which is cultivated in India for both oil production and the flower market, produces about 4,200 kilograms of flowers per hectare from April to August; whereas *J. grandiflorum* flowering from about May to August can yield up to 10,000 kilograms per hectare (Weiss, 1997).

OIL EXTRACTION

In earlier times, enfleurage was the traditional method of extraction of oil from the flowers. This method produces oil of a very high quality and it is very expensive, but still occasionally used and highly-prized by connoisseurs. It involves pressing flowers individually by hand into thin layers of pure fat (beef or pork fat was used in Europe) between plates, known as the chassis, for 24 to 48 hours. The spent flowers are then removed and fresh ones added. The process is repeated many times until the fat is saturated with the oil. The oil is then extracted with alcohol to create *jasmin pomade*. The spent flowers may be used to give a secondary extraction of a lower quality. This could be loosely compared to the cruder methods which have been used traditionally in south east Asia. Here perfumed oils were produced by repeatedly layering flowers between sesame seeds which are eventually crushed to release a jasmine-scented oil. Even the tradition of placing fresh flowers in oiled hair for fragrance has a similar result as the oil absorbs the flower oils.

Alternative extraction methods were required to make the production costs of jasmine oil reasonable. Many essential oils (such as attar of roses) are produced by steam extraction but this method cannot be used for jasmine as the delicate oil is adversely affected by the heat. Additionally, steam extraction produces a very low yield.

Solvent extraction is now used, whereby the fresh flowers are washed several times with hydrocarbon-type solvents such as high-grade hexane. This must be carried out near the jasmine fields to ensure that there is no change in the aromatic components, but the resulting 'concrete' may then be shipped elsewhere for further processing to produce the absolute.

The concrete is a semi-solid, waxy, yellow to yellowish-brown or orange substance which contains about 50% volatile oil. The absolute is obtained from concrete by a secondary process using ethanol. After extraction of oil, the remaining wax retains some scent and can be used for other products such as soaps. Approximately one ton of flowers yields 2.5 to 3 kilograms of concrete (Guenther, 1952).

The main producers, Egypt and Morocco, produce approximately 12–15 tons per year and the main markets are Western Europe, especially France, and the USA. The world trade and use of jasmine oil is extensive. In 1994 imports into the USA were 10,329 kilograms at $202.10 per kilogram, compared to rose oil with 5,443 kilograms at $1,233.44 and lavender oil with 339,621 kilograms at $14.67 (Weiss, 1997).

MEDICINAL USES

AROMATHERAPY

Jasmine oil is a popular choice in aromatherapy for a number of complaints ranging from nervous, respiratory, genito-urinary and muscular conditions. It is also used in skin care and is reputed to produce feelings of optimism and a lifting of the spirits.

TRADITIONAL MEDICINE

In almost all countries where jasmine grows naturally, there are records of traditional medicinal use. Tibet, China, Thailand, Nepal, Cuba, Cambodia, Vietnam, Fiji, Africa and others list it as playing an important role in complaints from snakebite to smallpox, from aphrodisiacs to antiseptics to antidepressants, from tetanus to tonics. All parts of the plants have been used; root, bark, foliage, flowers, fruit and the oil but it is the leaves which appear to be used most frequently. The main uses appear to be in treating fever, bronchial and respiratory disease, ringworm, ulcers, the suppression of milk production, intestinal problems, headaches, ulcers and for the blood and heart. Some scientific research has been carried out into the effectiveness of these treatments with some positive results. There are, however, a very small numbers of reports of poisoning by jasmine.

4. CULTIVATION:
USING JASMINES IN THE GARDEN

HISTORY

Jasmines have been long cultivated for their fragrance in the Far East, especially China and Arabia. The botanical name itself is derived from the ancient Persian name *yasmin* and the plant has been the inspiration of poets in Europe and Asia over the centuries. One of the first records in China appears to be in ancient Chinese texts where *Jasminum sambac* is recorded as having been introduced into that country near the end of the in the third century AD (Li, 1959). It was also very popular in Persia and other parts of Arabia, and possibly in Egypt in ancient days.

Clusius suggests that *Jasminum sambac* was introduced to Italy from Cairo around 1600 and later into England. It is so widely grown in the tropics and has been grown by so many ancient cultures that its true origin is not certain, although it is now considered to have originally come from India. Jasmine was mentioned by Dioscorides in the first century AD although which species is not known. In western Europe the earliest introductions of jasmines appear to be many centuries later, possibly one of the earliest being *Jasminum officinale* by the Moors in Spain between the eighth and eleventh centuries or even earlier along the long-established and ancient trade routes from Europe to the East. There are suggestions that the Portuguese navigator, Vasco da Gama knew about jasmines, although which species is uncertain, and he could have brought plants back from the East.

Although many of the dates of introduction are uncertain, jasmines were definitely used for their scent, as a source of perfumed oils, for medicinal purposes, in religious ceremonies and as ornamental plants in gardens around the world for at least 2000 years.

Several jasmines are mentioned in the early English herbals including *Jasminum officinale* by William Turner in *The Names of Herbes* in 1548 and others in Gerard's *Herbal* of the late sixteenth century, but they were the subject of mediaeval poetry and were probably known for many years before. Sir Thomas Hanmer, in his *Garden Book* (1659), grew *J. officinale* and *J. grandiflorum*, which he calls Spanish or Catalonian; it was grown in pots and pruned nearly to the ground each winter, the new shoots flowering when not much 'above a yard high'. One of these plants, treated as an herbaceous pot plant, is illustrated in *Hortus Eystettensis* (1613). Hanmer also reports that it was grown outside on south walls in Paris, covered with moss and horse dung in winter and with the branches coated in wax. He states that they seemed to survive the colder winters there better than in England. He also reports that sambac, called Arabian or Alexandrian Jasmyn, in Italian *gelsimeno del Gime*, formed a dwarf shrub; this appears to be the clone now called 'Grand Duke of Tuscany'. Two yellow-flowered species were also cultivated then; 'Yellow Indian', probably *J. humile,* and *J. fruticans* 'rather a broom than a jasmine, of little beauty and esteem'.

CULTIVATION

Although over 200 species of jasmine are recognised in the wild, only about 50 are now cultivated; and even of these, the majority are tender and therefore grown only in tropical or subtropical parts of the world. A few of the tender species are grown under glass in temperate regions and some may be suitable for growing in containers that can be protected during winter. Though fewer than ten are completely frost hardy they are not difficult to grow in cooler gardens and the majority of them are readily available in the trade. Details of any particular requirements are mentioned under the individual species.

A number of jasmine species are very vigorous. In the tropical regions, several jasmines grow wild in such abundance around the world that it has been difficult to ascertain the exact origin of the species. For example, *Jasminum fluminense* was initially described as a native of Brazil, but was subsequently found to grow wild in western Africa and was probably introduced to Brazil by early Portuguese traders. It is also naturalised in Hawai'i, but does not appear to be affecting the local flora. However this species and *J. dichotomum,* also from Tropical Africa, were introduced into gardens in Florida, USA and have adapted so well to the conditions there that they are now considered to be invasive pests and their spread is discouraged. *Jasminum laurifolium* is also considered a potential self-seeding weed. In temperate regions however, none have become a problem.

USING JASMINES IN THE GARDEN

All jasmine species are woody but they have different habits and can be climbing, scrambling or shrubby. The vigorous climbing species such as *Jasminum* × *stephanense* or *J. multiflorum* may be grown in a number of situations. It is important to provide some form of support around which the plants can twine, and when first planted it is usually necessary to tie the plant to the chosen support to encourage it to start to twine. Climbers can be used on a pergola or trellis, to grow up a simple pillar and to cover fences or walls. In some cases, they may be used as ground cover on a bank. In cooler climates, the less hardy species can be grown up supports in a conservatory or greenhouse, but the most vigorous species may need to be confined in smaller spaces. *J. polyanthum* is frequently grown as an indoor container plant in northern Europe and is often trained on a circular support.

The species with a more scrambling habit are less able to support themselves and will require more tying in, but they can also be used to cover a pergola, trellis, fence or pillar and will make ideal ground cover on a bank; with support and pruning they can even create informal hedges. *Jasminum nudiflorum* and *J. humile,* for example, are ideal in all these situations and have even been used for bonsai. By early training some naturally scrambling plants such as *J. molle* can be made into bushes.

The non-climbing species which naturally form shrubs, such as *Jasminum fruticans* or *J. leptophyllum,* are very versatile and may be planted in shrub or mixed borders, against the house or again used for hedging. The smallest species, *J. parkeri,* makes a useful addition in the rock garden or alpine house. It may be grown in a container, as a specimen plant or encouraged to form a low-growing ground cover.

All species are useful to wildlife; the flowers encourage butterflies and other insects, while the fruits will attract birds. The dense growth of unpruned jasmines is ideal as a nesting site.

GROWING IN TEMPERATE GARDENS

Out of doors, hardy species are easy and trouble-free plants. They survive in most good garden soils in sun or partial shade. The majority thrive in a well-drained soil although they should not be allowed to dry out completely. A sunny warm position is ideal as although most will tolerate shade, they will flower better in sun. *Jasminum nudiflorum* can be planted on a north facing aspect. Those requiring frost protection should be grown in a greenhouse or conservatory with good light in a good fertile soil, and the climbing species will require support. Many of those which are on the verge of frost hardiness, such as *J. polyanthum*, may be grown in a large pot and moved out of doors in the summer where they will benefit from the additional sun to ripen the wood. Plants grown in containers will require regular extra feeding, watering and repotting; on the other hand, a container will restrict the roots and hence the shoots meaning more vigorous plants may be grown in a smaller area. Regular pruning will be necessary to keep the more vigorous species under control.

GROWING IN TROPICAL AND SUBTROPICAL GARDENS

Frost tender species growing in subtropical and tropical climates are generally trouble free, and given sufficient water and fertile soil will make vigorous plants. In gardens in tropical areas, the temperate species thrive at higher altitudes and need watering if the climate is too dry.

SIZE AND VARIATION OF JASMINE SPECIES

The sizes of plants in the descriptions below give the range found both wild and cultivated plants. However, it must be remembered that the growing conditions including soil, water, altitude and temperature, whether in a garden, conservatory or in the wild can have an important impact on the vigour of a plant. Heights of climbers have been omitted because the height to which a strong growing and vigorous plant can reach is very variable and will depend to an extent on the height of the support and the conditions in which it is growing. For instance, plants grown in a container or glasshouse will tend to be smaller than those grown in the open ground. The heights of shrubby plants and those with a tendency to scramble have been included.

It should also be remembered that in the wild, plants show considerable natural variation in many characteristics such as size, hardiness and even fragrance, both within a population and in different natural geographical locations. The majority of species, at least in western Europe and North America, are the result of a single introduction, perhaps as seed or as a living plant, from a single location in the wild. As these have since been vegetatively propagated for distribution commercially, the plants with which we are familiar in cultivation may be very different in appearance to newer introductions from a different area or population. This has sometimes given rise to confusion in identity and nomenclature and numerous new and unnecessary specific names which have been confused in the nursery trade.

Flowering times may also be different under cultivation and can be deliberately altered for commercial purposes, especially when grown for the pot plant trade.

PRUNING AND TRAINING

Most jasmines climb by twining around a support; others have a scrambling habit or a loose arching habit which if allowed to grow unchecked will create a mounded plant with a build-up of dead wood within. Others are shrubs and pruning is carried out to encourage side branches and form a bushier, well-shaped plant. How the plants are trained and the pruning regime will be partly determined by their habit; other important points to consider are how the plants will be used in the garden, and whether the species flower on the current or previous year's wood. As a general rule, pruning would normally be carried out soon after flowering, unless the fruits are specifically needed, so that those flowering early in the year on the previous year's wood are pruned in spring or early summer whereas those flowering on the current season's growth will be pruned later in the season. There are of course exceptions and for those tropical species that tend to flower intermittently all the year around, this general rule cannot be easily applied.

SHRUBS

In general, the hardy shrubby species tend to flower on the current season's wood. Cut away flowering branches and any weak or dead wood after flowering as necessary to maintain the required size and shape.

CLIMBERS

The vigorous climbers, both hardy and tender, may need restrictive pruning after flowering, but in general, if encouraged initially, climbers will climb naturally and need little pruning. In a greenhouse or conservatory, if the plants are reaching for the light and become straggly below while flowering towards the top of the plant, reducing the height may make the flowers more visible and the plants bushier and more attractive. Again, prune after flowering. In tropical or subtropical climates, climbers will need little attention.

SCRAMBLERS

These benefit from pruning the long shoots from an early stage to avoid them becoming too straggly and to encourage a bushy rather than a straggling plant. Left alone scramblers will climb into themselves to form a sprawling bush. If this is what is required or if the plant is to be used as ground cover, simple pruning to keep it within bounds is all that is necessary. *Jasminum nudiflorum* and *J. mesneyi* flower early in the season on the previous year's wood and should be trimmed after flowering in late spring.

HEDGES

A species to be trained as a hedge will need initial support until a firm framework is established and thereafter regular pruning after flowering.

PROPAGATION

If fertile fruits are produced, propagation of species is easiest by sowing the seed when fresh, but because many plants in cultivation are all derived from a single clone, vegetative propagation is the most reliable method. Because many, if not all, species are self-incompatible, the most fertile seed is produced by cross-pollination, and seed from self-pollination is likely to be weak or infertile.

Many species will naturally layer themselves so that when the branches touch the ground, roots are formed. These species, including *Jasminum beesianum* and *J. angulare*, are easy to layer by deliberately pegging flexible shoots to the ground in spring in a fertile free draining soil or compost to encourage rooting. A good root system will be formed in less than 12 months after which the rooted section should be carefully detached from the parent plant and potted up into a standard compost and planted out once well established. Avoid detaching the young plant too early but wait until the roots are well established and shoots growing strongly.

All jasmines may be propagated quite easily by cuttings at almost any time of year, although some species are slower to root than others. In spring or early summer, internodal or nodal softwood cuttings may be taken before the wood begins to harden. Cuttings of about 5 cm long, bearing at least one pair of leaves and cut beneath a node (leaf joint), normally give the best results. Really soft growth should be removed. If the leaves are large, they may be reduced in size to avoid wilting or the development of mould before the roots are developed. The use of rooting hormone will aid the rooting process as will mist and bottom heat. Rooting will take about one month. Once well rooted, the young plants may be potted into suitable compost and kept in a sheltered position until well established and ready for planting in their final position in the ground. Semi-ripe cuttings may also be taken in the same way as the wood begins to ripen in summer. In temperate climates, it is advisable to aim to get the cuttings rooted and the plants established by midsummer so that they

Fig. 5. **1.** Long shoot of *Jasminum*, suitable for cutting material. **2.** Cutting with one node. **3.** Cutting with three nodes.

Fig. 6. Rooted cutting.

are able to harden off and survive the colder winters. In a very dry climate, misting will help to increase the humidity in the atmosphere. Hardwood nodal (heel) cuttings of the hardy species, taken in early winter, root best in free draining compost and in a cool sheltered position in a cold frame or frost-free greenhouse, to avoid drying by cold winds. Do not allow the cuttings to dry out, but equally do not let them become waterlogged. The cuttings should be rooted by spring, but give them time to become well established before potting up and growing on until ready for their final planting out. Frost tender species will require more heat and are probably best not propagated by this method except in tropical climates.

An interesting note made by Cibot in 1778 is mentioned by Peter Valder in *The Garden Plants of China* (1999). It gives an insight into propagation in China about 300 years ago and explains how cuttings were pushed through half-rotted boards floating on water and kept in the shade. When the cuttings were rooted, the boards were broken away and the young plants planted in soil.

PESTS, DISEASES AND DISORDERS

In general there are few pests and diseases which are specific to jasmine species, and being tolerant of a wide range of conditions they are rarely affected by poor or unsuitable soil conditions. In temperate countries, as long as the plants are not waterlogged and are grown in a sunny position, the hardy species are not difficult to grow. For the more tender species, grown under greenhouse or conservatory conditions, too much heat or intense light may scorch the young foliage. Otherwise, regular checks, and treatment if required, should be made for the usual pests such as red spider mite, white fly and scale insect before they become problematic. Excess humidity may result in fungal attack especially in winter when the temperatures are cooler in northern Europe. Aiming to keep the air buoyant by ample ventilation will help alleviate the problem. The usual sensible precautions of horticulture should be observed and if a plant does become affected by any serious problem and needs to be destroyed, avoid replanting a replacement of the same species in the same position. However, jasmines on the whole appear to be very disease and pest free. The main problems with pests and diseases will occur when the plants are grown on an intensive scale for the commercial production of flowers for oil or leis. Under these conditions, a pest or disease can multiply rapidly and there have occasionally been examples of crops being totally lost.

HARDINESS

The hardiness quoted below for each species follows the *European Garden Flora* ratings, and their approximate USDA zone equivalents.

These zones are approximations and hardiness depends on a number of other factors including length of time below a certain temperature, late or early frosts and humidity. As many Jasmines are grown against walls, they will usually get a degree of frost-protection overnight, though there is less effect during long spells of frosty weather.

The hardiness ratings and zones are as follows:

H1 20°C or less: (will survive almost anywhere but no jasmines fall in this category), more or less equivalent to USDA Zone 5.

H2 -15 to -20°C: more or less equivalent to USDA Zone 6.

H3 -10 to -15°C: includes most of the UK, more or less equivalent to USDA Zone 7.

H4 -5 to -10°C: includes the UK except for high areas or frost spots, more or less equivalent to USDA Zone 8.

H5 0 to -5°C: survives in the mildest sheltered gardens, more or less equivalent to USDA Zone 9.

G1 above 0°C: may tolerate short slight frosts but normally requires cool greenhouse even in southern Europe, more or less equivalent to USDA Zone 10.

G2: needs heated greenhouse even in southern Europe, more or less equivalent to USDA Zone 11.

CHOOSING WHICH JASMINES TO GROW

In this alphabetical table the main characters of the species can be seen at a glance.

Note that flowering times are those of plants growing in the wild. Many of the tropical and subtropical species will flower intermittently throughout the year. In cultivation this may be different depending on the conditions in which the plants are grown.

Plant	Hardiness	Habit	Flower Colour	Scent	Flowering Time	Other points
J. × stephanense	H4	Climber	Pink	Yes	Summer	
J. 'Ann Clements' (*J. multiflorum* × *J. nobile* subsp. *rex*)	G2	Climber	White	No	Intermittently all year round	
J. abyssinicum	G2	Climber	White sometimes flushed pink outside	Yes	Winter to early spring	
J. adenophyllum	G2	Climber	White	Yes	Spring to late summer	
J. angulare	G1	Climber	White tinged pink in bud	Yes	Summer	
J. angulare 'Anne Shelton'	G1-H5	Climber	White sometimes flushed pink outside	Yes	Late summer and later into autumn or winter	Slightly hardier than species
J. auriculatum	G2	Climber	White	Yes	Summer	
J. azoricum	G1-2	Weak	White	Yes, strong	Winter to autumn depending on conditions	
J. beesianum	H3	Weak climber	Red to deep rose-pink	Yes	Late spring to summer	
J. bignoniaceum	H5-G1	Shrub	Yellow	No	Intermittent from spring to early autumn	
J. decussatum	G2	Climber	White	Yes	Winter to spring	
J. dichotomum	G2	Climber	White tinged red outside	Yes, sweet scent	More or less all year round	
J. didymum subsp. *didymum*	G2	Climber	White to creamy white	Yes	Winter	
J. didymum subsp. *lineare*	G2	Climber	White	Yes	Winter	
J. didymum subsp. *racemosum*	G2	Climber	White	Yes	Winter	
J. dispermum	H5–G1	Climber	White sometimes tinged pink	Yes	Late spring to summer	

Plant	Hardiness	Habit	Flower Colour	Scent	Flowering Time	Other points
J. elongatum	G1–2	Climber or scrambling shrub	White	Yes	Spring to summer	
J. flexile	G1–2	Climber	White	Yes	Winter to spring	
J. floridum	H3	Shrub	Yellow	Yes	Summer to autumn	
J. fluminense	G2	Climber or scrambling shrub	White, bud tinged pink	Yes	Spring to summer	
J. fruticans	H4	Shrub	Yellow	Yes	Summer	
J. grandiflorum	G1	Weak climber	White sometimes tinged pink	Yes, strong	Mainly summer	
J. grandiflorum 'De Grasse'	G1	Weak climber	White sometimes tinged pink	Yes, strong	Mainly summer	
J. grandiflorum subsp. *floribundum*	G1	Weak climber	White sometimes tinged pink	Yes, strong	Mainly summer	
J. harmandianum	G2	Climber	White	Yes	Spring to summer	
J. humile	H5	Shrub sometimes somewhat scrambling	Yellow	Some	Late spring to summer	
J. humile 'Revolutum'	H5	Shrub	Yellow	Some	Late spring to summer	Larger than species
J. humile f. *farreri*	H5	Shrub	Yellow	Some	Late spring to summer	Slightly downy
J. humile f. *wallichianum*	H5	Shrub sometimes somewhat scrambling	Yellow	Some	Late spring to summer	More leaflets
J. kedahense	G2	Climber	White	Yes	Winter to spring	
J. lanceolaria	G2	Climber	White	Yes, sweet scent	Spring to summer	
J. laurifolium f. *nitidum*	G2	Climber but initially more shrub-like	White sometimes tinged pink	Yes	Late spring to autumn	
J. leptophyllum	H4	Shrub	Yellow	Yes	Summer	Very narrow leaves; tolerant of drought
J. maingayi	G2	Climber	White	Yes	Autumn to spring	

Plant	Hardiness	Habit	Flower Colour	Scent	Flowering Time	Other points
J. mesneyi	G1–H5	Scrambling shrub	Yellow	No	Spring	Double flowers
J. molle	G2	Climber or weak climber	White	Yes	Spring to summer	
J. multiflorum	G2	Weak climber	White	Yes, strong	Winter to spring	
J. multipartitum	G1	Weak climber	White some pink outside	Yes	Spring to summer	Large flower
J. nobile subsp. *rex*	G2	Weak climber	White some pink outside	No	More or less all year round	Largest flower of genus
J. nudiflorum	H2	Scrambling shrub	Yellow	No	Winter to early spring	Deciduous
J. nudiflorum 'Aureum'	H2	Scrambling shrub	Yellow	No	Winter to early spring	Deciduous; yellow variegated leaves
J. nudiflorum 'Mystique'	H2	Scrambling shrub	Yellow	No	Winter to early spring	Deciduous; white edged leaves
J. odoratissimum	G1	Shrub or scrambling shrub	Yellow	Variable	Summer	
J. officinale	H4	Climber	White some pink outside	Yes	Summer	
J. officinale 'Argenteovariegatum'	H4	Climber	White some pink outside	Yes	Summer	Leaves edged white
J. officinale 'Aureum'	H4	Climber	White some pink outside	Yes	Summer	Foliage yellow
J. officinale 'Clotted Cream'	H4	Climber	Cream	Yes	Summer	
J. officinale 'Crûg's Collection'	H4	Climber	White some pink outside	Yes	Summer	
J. officinale 'Devon Cream'	H4	Climber	Cream	Yes	Summer	
J. officinale 'Inverleith'	H4	Climber	White red outside	Yes	Summer	
J. officinale f. *affine*	H4	Climber	White some pink outside	Yes	Summer	
J. officinale FIONA SUNRISE 'Frojas'	H4	Climber	White some pink outside	Yes	Summer	Yellow foliage
J. parkeri	H4	Dwarf shrub	Yellow	Yes	Early summer	Very small plant

Plant	Hardiness	Habit	Flower Colour	Scent	Flowering Time	Other points
J. polyanthum	H5	Climber	White pink outside	Strong	Spring or winter when forced for pot plant trade	Sometimes flowers again in late summer
J. sambac	G2	Climber or suberect shrub	White	Very strong	More or less all year round	
J. sambac 'Asian Temple'	G2	Shrub	White	Very strong	More or less all year round	Double flowers
J. sambac 'Bangkok Peony'	G2	Shrub	White double	Very strong	More or less all year round	Loosely double flower
J. sambac 'Belle of India'	G2	Shrub to weak climber	White	Very strong	More or less all year round	Double flower with twisted petals
J. sambac 'Grand Duke of Tuscany'	G2	Shrub	White	Very strong	More or less all year round	Very double almost spherical flower
J. sambac 'Little Bo'	G2	Shrub	White	Very strong	More or less all year round	Dwarf plant with double flower
J. sambac 'Maid of Orléans'	G2	Shrub	White	Very strong	More or less all year round	Semi-double flower
J. sambac 'Thai Beauty'	G2	Shrub	White	Very strong	More or less all year round	Hose in hose double flower
J. scandens	G2	Climber	White	Yes	Winter to spring	
J. simplicifolium subsp. *australiense*	G1	Climber	White	Yes	Spring to summer	
J. simplicifolium subsp. *leratii*	G1	Weak climber	White	Variable	Spring to summer	
J. simplicifolium subsp. *suavissimum*	G1	Weak	White	Yes	Spring to summer	
J. sinense	G2	Climber	White tinged cream	Yes	Intermittently through the year	
J. subhumile	G2	Shrub	Yellow	Yes	Late spring to summer	
J. syringifolium	G2	Climber	White	Yes	Winter to spring	
J. tortuosum	G1	Climber or scrambling plant	White	Yes	Summer	
J. urophyllum	G2	Climber	White with faint pink flush outside	Yes	Late spring to summer	

5. TAXONOMIC TREATMENT

SYNOPSIS OF THE GENUS JASMINUM

SECTION JASMINUM

Leaves opposite, usually pinnate, rarely simple or trifoliate; not articulate. Flowers often flushed reddish-purple on the outside. Seeds one or two in each fruit. Type: *J. officinale*. Species range from south western and western China through the Himalayas and India to the Caucasus and Arabia (including northeastern Africa).

1. *Jasminum officinale*
  cv. 'Argenteovariegatum'
  cv. 'Inverleith'
  f. *affine*
  cv. 'Aureum'
  cv. 'Devon Cream'
  cv. 'Clotted Cream'
  cv. 'Crûg's Collection'
  cv. FIONA SUNRISE 'Frojas'
  cv. 'Grandiflorum'

2. *Jasminum grandiflorum*
  2a. subsp. *grandiflorum*
  2b. subsp. *floribundum*

3. *Jasminum polyanthum*

4. *Jasminum dispermum*

5. *Jasminum beesianum*

6. *Jasminum* ×*stephanense*

SECTION TRIFOLIOLATA DC.

Leaves opposite, usually trifoliolate; petioles not articulate. Flowers sometimes small. Usually with one seed per fruit except *J. abyssinicum* which is closer to Section Jasminum in seed characters. Type: *J. didymum*. The species range from southern, eastern and tropical Africa, Madeira, India, the Himalayas south western China and Australia.

7. *Jasminum abyssinicum*

8. *Jasminum angulare*
  cv. 'Anne Shelton'

9. *Jasminum auriculatum*

10. *Jasminum azoricum*

11. *Jasminum didymum*
  11a. subsp. *didymum*
  11b. subsp. *lineare*
  11c. subsp. *racemosum*

12. *Jasminum flexile*

13. *Jasminum fluminense*

14. *Jasminum lanceolaria*

15. *Jasminum sinense*

16. *Jasminum tortuosum*

17. *Jasminum urophyllum*

SECTION UNIFOLIOLATA DC.

Leaves opposite, usually simple, petioles articulate. Usually with one seed per fruit except *J. simplicifolium* which has two. Type: *J. sambac*. The species are mainly found around the Pacific from Australia, Tropical Africa, and Tropical Asia including Thailand, Malaysia, Cambodia, Laos, Burma (Myanmar), India and China.

18. *Jasminum adenophyllum*

19. *Jasminum decussatum*

20. *Jasminum dichotomum*

21. *Jasminum elongatum*
    21a. *Jasminum perissanthum*

22. *Jasminum harmandianum*

23. *Jasminum kedahense*

24. *Jasminum laurifolium* forma *nitidum*

25. *Jasminum maingayi*

26. *Jasminum molle*
    26a. *Jasminum calcareum*
    26b. *Jasminum kajewskii*

27. *Jasminum multiflorum*

28. *Jasminum multipartitum*
    28a. *Jasminum glaucum*

28b. *Jasminum streptopus* var. *transvaalensis*
28c. *Jasminum stenolobum*
29. *Jasminum nobile* subsp. *rex*

30. *Jasminum sambac*
    cv. 'Asian Temple'
    cv. 'Bangkok Peony'
    cv. 'Belle of India'
    cv. 'Grand Duke of Tuscany'
    cv. 'Little Bo'
    cv. 'Maid of Orlèans'
    cv. 'Thai Beauty'

31. *Jasminum scandens*

32. *Jasminum simplicifolium*
    32a. *J. simplicifolium* subsp. *suavissimum*
    32b. *J. simplicifolium* subsp. *australiense*
    32c. *J. simplicifolium* subsp. *leratii*

33. *Jasminum syringifolium*
    33a. *Jasminum duclouxii*

SECTION ALTERNIFOLIA DC.

Leaves alternate; petioles not articulate. Flowers yellow. Plant rarely climbing. Two seeds in each fruit. Type: *J. humile*. Species range from northern and central China through India, Pakistan, Nepal, Afghanistan, Iran and parts of southern Europe with a few outlying species found in Madeira and the Canary Islands and East Africa.

34. *Jasminum bignoniaceum*

35. *Jasminum floridum*

36. *Jasminum fruticans*

37. *Jasminum humile*
    cv. 'Revolutum'
    37a. forma *wallichianum*
    37b. forma *farreri*

38. *Jasminum leptophyllum*

39. *Jasminum odoratissimum*

40. *Jasminum parkeri*

41. *Jasminum subhumile*

SECTION PRIMULINA P. S. Green

Leaves trifoliolate, opposite; petioles not articulate. Flowers yellow. Type: *J. nudiflorum.* Species come from Western China. A small section with only two species.

42. *Jasminum mesneyi*

43. *Jasminum nudiflorum*

KEY TO THE CULTIVATED SPECIES OF JASMINUM

Please note that *J. simplicifolium* is not included in the key as it is probably not widely cultivated. Its cultivated subspecies are included. Other subspecies can be identified by keys within the main species account. Related species, cultivars, varieties and forms are not included in the key.

1.   Leaves alternate; flowers yellow 2
1a. Leaves opposite; flowers yellow, white or red 9

2.   Prostrate or low-growing shrub rarely exceeding 0.3 m; leaves *c.*1 cm . . . . . **40. *J. parkeri***
2a. More or less erect shrub to 5 m; leaves longer than 1 cm 3

3.   All leaves simple, linear to linear-lanceolate . . . . **38. *J. leptophyllum***
3a. The majority of leaves pinnate or trifoliate although a few simple leaves may also be present 4

4.   Calyx-lobes as long or longer than the calyx tube, usually 2.5 mm or more 5
4a   Calyx-lobes shorter than the calyx tube, usually under 1.5 mm 6

5.   Inflorescence 6- or more-flowered; terminal leaflet elliptic or narrower . . . **35. *J. floridum***
5a. Inflorescence 1–4-flowered; terminal leaflet broadly elliptic . . . . **36. *J. fruticans***

6.   Inflorescence 20–120-flowered, not umbellate . . . . **41. *J. subhumile***
6a. Inflorescence under 25-flowered 7

7.   Inflorescence under 6-flowered; corolla tube funnel-shaped, widening towards the mouth . . . . **34. *J. bignoniaceum***
7a. Inflorescence usually over 6-flowered; corolla tube not funnel-shaped, not widening towards the mouth 8

8.   Leaflets usually 5 or more; inflorescence more or less umbellate, with flowers appearing more or less at same level; pedicels usually exceeding 1 cm . . . . **37. *J. humile***
8a. Leaflets usually 3; inflorescence not more or less umbellate, with flowers appearing at different levels; pedicels usually under 1 cm . . . . **39. *J. odoratissimum***

9.   Flowers yellow 10
9a. Flowers not yellow 11

10. Leaves deciduous, appearing after flowers; flowers 2–2.5 cm across, petals 6 . . . . **43. *J. nudiflorum***
10a. Leaves persistent, present at flowering; flowers 3–4.5 cm across, petals 6 or more . . . . **42. *J. mesneyi***

11.   Leaves with more than one leaflet; if of one leaflet, flowers pink or red 12
11a.   Leaves with one leaflet; flowers always white or white flushed with pink or red but
       never totally red or pink 28

12.   Corolla red or pink inside as well as outside 13
12a.   Corolla white inside but sometimes flushed reddish purple outside 14

13.   Leaves always unifoliolate . . . .**5. J. beesianum**
13a.   Leaves with variable number of leaflets, usually 3–5 . . . . **6. J. ×stephanense**

14.   Leaves with more than 3 leaflets, arranged pinnately 15
14a.   Leaves with 3 leaflets; if more, leaves digitate not pinnate 18

15.   Calyx lobes less than 2 mm, rarely linear 16
15a.   Calyx lobes linear, over 2 mm and usually 5–8 mm 17

16.   Leaflets 3 veined; corolla tube longer than 20 mm; flowers mainly in winter to spring . . . .
       . . . . **3. J. polyanthum**
16a.   Leaflets pinnate veined; corolla tube less than 15 mm long; flowers mainly in spring
       and summer . . . . **4. J. dispermum**

17.   Inflorescence more or less umbellate; corolla lobes usually under 12 mm long; hardy
       in most areas of Britain . . . . **1. J. officinale**
17a.   Inflorescence with middle pedicels conspicuously shorter than lateral ones; corolla
       lobes usually over 12 mm long; barely hardy even in mildest parts of Britain . . . .
       . . . . **2. J. grandiflorum**

18.   Calyx lobes minute or not exceeding 2 mm 19
18a.   Calyx lobes exceeding 2 mm, usually equal to or exceeding calyx tube 27

19.   Corolla usually exceeding 2 cm in diameter 20
19a.   Corolla rarely exceeding 1 cm in diameter . . . . **11. J. didymum**

20.   Leaves usually grey pubescent, lateral leaflets often very small or absent . . **9. J. auriculatum**
20a.   Leaves glabrous except for possible tufts of hairs in veins axils below; if pubescent
       below, lateral leaflets well developed 21

21.   Inflorescence with up to 5 flowers 22
21a.   Inflorescence with more than 5 flowers 23

22.   Leaflets oblong, ovate or broader and with tufts of hairs in axils of veins . . . **8. J. angulare**
22a   Leaflets linear or lanceolate but without tufts of hairs in axils of veins . . . **16. J. tortuosum**

23.   Leaflets very leathery; no tufts of hairs in axils of veins; inflorescence with dense
       clusters of flowers on short pedicels . . . . **14. J. lanceolaria**
23a.   Plants without this combination of characters 24

24.   Pedicels to 10 mm or more 26
24a.   Pedicels always under 7 mm 25

25.   Leaflets consistently with clearly visible small tufts of hairs in axils of veins . . . **12. J. flexile**
25a.   Leaflets sometimes with indistinct tufts of hairs in axils of veins . . . . **13. J. fluminense**

26.   Leaflets with clear tufts of hairs in axils of veins . . . . **7. J. abyssinicum**
26a.   Leaflets without clear tufts of hairs in axils of veins . . . . **10. J. azoricum**

27.   Leaflets with 3 very clearly defined veins, sometime slightly hairy . . . . . **17. *J. urophyllum***
27a. Leaflets with 4–5 pairs of somewhat obscure veins, very hairy especially on underside . . . . .
. . . . **15. *J. sinense***

28.   Calyx lobes equal to or shorter than calyx tube 29
28a. Calyx lobes exceeding calyx tube and usually 5 mm or more long 36

29.   Plant always hairy 30
29a. Plant glabrous or only minutely hairy 31

30.   Pedicels 1–20 mm; inflorescence open; corolla lobes 5–6 . . . . **26. *J. molle***
30a. Pedicels 0–3 mm; inflorescence very congested; corolla lobes 7–8 . . . . . .**19. *J. decussatum***

31.   Inflorescence dense; flowers with short pedicels under 3 mm; leaves leathery with
      3 very clear deeply impressed veins . . . . **20. *J. dichotomum***
31a. Plants without the above combination of characters 32

32.   Flowers small, under 2 cm, usually *c*.1 cm, across; pedicels 1–3 mm; inflorescence
      compact . . . . **31. *J. scandens***
32a. Flowers 1 cm or more across; pedicels usually over 3 mm and inflorescence open 33

33.   Leaves consistently very narrow, *c*.2–3 mm wide, never over 10 mm wide . . . .
      . . . . **32a. *J. simplicifolium*** subsp. ***suavissimum***
33a. Leaves broader, over 10 mm wide, if narrower, not consistently so 34

34.   Lower pair of leaf veins not forming an obscure marginal vein . . . . . . **33. *J. syringifolium***
34a. Lower pair of leaf veins forming an obscure marginal vein 35

35.   Corolla tube to 20 mm, rarely under 12 mm . . . . **32c. *J. simplicifolium*** subsp. ***leratii***
35a. Corolla tube under 12 mm . . . . **32b. *J. simplicifolium*** subsp. ***australiense***

36.   Flowers not fragrant; inflorescence 1–3 flowered; flowers to 6 cm across . . . .
      . . . . **29. *J. nobile*** subsp. ***rex***
36a   Flowers fragrant; smaller 37

37.   Calyx lobes held at right angles to the calyx tube . . . . **24. *J. laurifolium*** f. ***nitidum***
37a. Calyx lobes not held at right angles to the calyx tube 38

38.   Pedicels usually very long to 4 cm; leaves with conspicuous tufts of hairs in the axils
      of the veins . . . . **18. *J. adenophyllum***
38a. Pedicels not exceeding 1 cm; leaves without conspicuous tufts of hairs in the axils of
      the veins, or if present not consistent or conspicuous to naked eye 39

39.   Flowers solitary, sometimes to 3, in inflorescence; leaves usually not exceeding 3.5 cm
      long . . . . **28. *J. multipartitum***
39a. Flowers rarely solitary and if so, leaves exceeding 3.5 cm long 40

40.   Plant densely hairy; inflorescence congested with large leafy bracts to 2 cm . . . .
      . . . . **27. *J. multiflorum***
40a. Plant not densely hairy, if hairy without leafy bracts and inflorescence lax 41

41.   Corolla lobes broad, 5–9 mm wide; petioles short, 2–6 mm long . . . . **30. *J. sambac***
41a. Corolla lobes narrower, under 5 mm wide, if over 6 mm wide, petiole greater than
      6 mm long 42

42. Leaves not leathery or only slightly so 43

42a. Leaves leathery 44

43. Inflorescence crowded; pedicels 0.5–3 mm long; corolla lobes greater than 2 mm wide . . . . **21. J. elongatum**

43a. Inflorescence very lax; pedicels to 10 mm long; corolla lobes very narrow not exceeding 2 mm wide . . . . **22. J. harmandianum**

44. Leaves broad ovate to ovate or elliptic with lower veins forming a submarginal vein; corolla tube 25–35 mm long . . . . **23. J. kedahense**

44a. Leaves narrow ovate to oblong-ovate without a submarginal vein; corolla tube 15–30 mm long . . . . **25. J. maingayi**

THE GENUS JASMINUM

Section JASMINUM (Syn. Sect. *Pinnatifolia* DC.)

Section Jasminum is characterised by its opposite and pinnate leaves. It is a relatively small group, with 6 species and one hybrid, and includes the common jasmine, *Jasminum officinale*, the most familiar (and type) species of the genus. This section is sometimes called Section Pinnatifolia DC.

1. JASMINUM OFFICINALE

Section Jasminum

This species is probably the first jasmine to be grown in English gardens. It was recorded by William Turner in his book *The Names of Herbes* in 1548 and had probably been in cultivation for many years before being brought across Europe by traders along the ancient trade routes from China. It has a range of common names including common white jasmine, poets' jasmine, and jessamine, and is one of the jasmines of literature. Both white, 'Argenteovariegatum', and yellow-variegated, 'Aureum', leaved forms were already grown in 1787 at the time of the account in *Botanical Magazine* but they were noted to be less hardy.

It may be recognised by being a hardy white summer-flowering climber with pinnate leaves and long calyx lobes. It should not be confused with the more tender *Jasminum grandiflorum* in which the flowers are not borne in an umbel-like inflorescence.

In the wild, it is a very variable and widespread species, and towards the eastern end of its distribution, plants tend to have smaller leaves. It has become naturalised in parts of southern Europe and Asia.

In colder areas it may be deciduous and needs the protection of a south facing wall and training and support. Despite this, it is very reliable and easy to grow in a sunny position in well drained soil. It is a vigorous plant which will spread itself by layering and if the vegetative growth is too rampant, it may have fewer flowers.

Jasminum officinale L., *Sp. Pl.* 1: 7 (1753). Type: 'Habitat in India'. Herb. Clifford: Jasmine No. 1.

ILLUSTRATIONS. *Curtis's Botanical Magazine* 1: 31 (1787); Menninger, *Flowering Vines of the World*, p. 244, f. 198 (1970); *The Plantsman* 10 (3): 155 (1988); Brickell, *The RHS Gardeners' Encyclopedia of Plants & Flowers*, p. 166 (1989); Dirr, *Dirr's Trees and Shrubs for Warm Climates*, p. 158 (2002).

DESCRIPTION. Evergreen or semi-evergreen, more or less glabrous, climber. *Branches* ridged, green. *Leaves* opposite, pinnate, with 5–9 leaflets, the upper pair often sessile, petiole 5–15 mm long. *Leaflets* ovate to elliptic, slightly hairy below, with 2–4 obscure pairs of veins, below; base cuneate; apex acute to long acuminate; terminal leaflet much larger, 0.5–4.5 × 0.22 cm; lateral leaflets sometimes more rounded, 0.3–3 × 2–13 cm. *Inflorescence* terminal, more or less umbellate, 5 to 12-flowered, pedicels to 2 cm. *Bracts* linear, *c.*5 mm. *Flowers* white, sometimes tinged with pink or red on the outside or in bud, fragrant, 1.5–2.5 cm diameter. *Calyx* sparsely hairy, tube 1–3 mm long; lobes linear. 3–10 mm. *Corolla* tube, 1–2 cm, lobes 4–5, narrowly ovate to oblong, 9–11 mm. *Berries* purple, globose or ellipsoid, 7–10 × 5–9 mm.

Fig. 7. *Jasminum officinale*.

DISTRIBUTION. From the Caucasus through the Himalayas to SW China.
HABITAT. Valleys, ravines, thickets, woods, along rivers, meadows; 1,800–4,000 m.
FLOWERING TIME. Summer.

forma **affine** (Lindl.) Rehder
Syn. 'Grandiflorum'.
Flowers larger and usually with a pink tinge. This should not be confused with the tender species
J. grandiflorum (p. 54).

There are several cultivars, some ancient, some very recent. Those in cultivation at present are listed
below:

'Argenteovariegatum'
Syn. 'Variegatum'.
ILLUSTRATION. Brickell, *The RHS Gardeners' Encyclopedia of Plants & Flowers*, p. 166 (1989).
Leaves grey-green with creamy-white margins. This cultivar has been known since the eighteenth
century. Hardiness: H4.

'Aureum'
Syn. 'Aureovariegatum'.
ILLUSTRATION. Ellison, *Cultivated Plants of the World*, p. 329 (1995).

Plate 4. *Jasminum officinale* **'Inverleith'**. Painted by Christabel King for *Curtis's Botanical Magazine*, t.196 (1992).

Fig. 8. ***Jasminum officinale*** f. ***affine***. Habitat near Kanding, Sichuan.

Fig. 9. *Jasminum officinale* f. *affine*, near Kanding.

An old cultivar which has been known in gardens for at least 150 years, with leaves variegated and tinged yellow.

'Clotted Cream'
Very similar to 'Devon Cream', having been selected as a sport and introduced into cultivation in the United Kingdom in 2004.

'Crûg's Collection'
Collected from Mt Phulchoki, Nepal in 1995 by Bleddyn and Sue Wynn-Jones (*BSWJ* 2987). This selection has clusters of very fragrant, larger than average white flowers, tinged bright rose-pink, from summer to autumn. Plants reach a height of 8 m.

'Devon Cream'
Found by C. Parker as a chance seedling in Devon in 1994. It has cream coloured flowers which are slightly larger and more fragrant than the species. It was introduced into British gardens in 1998.

FIONA SUNRISE 'Frojas'
Leaves flushed golden yellow; found by Dave West as a seedling and introduced in 1995. At the time of writing, this cultivar is protected by Plant Breeder's Rights.

'Grandiflorum'
This has been used as a cultivar name, but most of these records refer to the species *Jasminum grandiflorum*. In Sweet's *Hortus Britannicus* (1827), a double-flowered form is listed and this is possibly still available today in North America.

Fig. 10 (left). *Jasminum officinale* **'Argenteovariegatum'**.
Fig. 11 (right). *Jasminum officinale* **Fiona Sunrise**.

'Inverleith'

ILLUSTRATION. *Kew Magazine* 9: 196 (1992).

A self-fertile and very floriferous cultivar, with flowers bright purplish-red in bud and on outside of flowers, when open contrasting with the pure white inside of petals. A similar plant was known at Glasnevin which had been received from France over 100 years ago, possibly via missionaries from SW China. It was almost certainly introduced from Yunnan to the Royal Botanic Garden, Edinburgh from Bodnant in 1953 as seed from the T. T. Yu expedition, (probably *T. T. Yu* 10586), but this has to remain hypothetical as no firm records are available. The photograph on p. 52, taken in western China near Kanding (Tachienlu), belongs to a very similar variety. Hardiness: H4.

2. JASMINUM GRANDIFLORUM

Section Jasminum

This jasmine is widespread as a cultivated plant in warm temperate and subtropical regions and is naturalised in some areas. It was brought into cultivation early on, by the Arabs, and introduced by them to Spain and other parts of Mediterranean Europe around the eighth to eleventh century. It has been known in England since at least 1629 when it was mentioned by Parkinson in *Paradisi in*

Sole and was often known as Spanish or Catalonian jasmine. It was sometimes grafted and in 1876, in the account in *Botanical Register*, grafted plants were considered to be hardier than those propagated by layering. It was also noted that the Moors used its woody stems for tobacco pipes by removing the pith. This species has one of the best scents in the genus, and is grown in southern France for the production of jasmine oil for the perfume industry.

Jasminum grandiflorum is a tender white-flowered scrambling plant with a very strong fragrance, distinguished from *J. officinale* by its larger flowers and less umbel-like inflorescence. The two have often been confused, so not all records in literature are reliable. In flower, *J. officinale* is easily distinguished by the character of the inflorescence in which the individual flowers are borne at more or less the same level, giving it an umbellate appearance, whereas in *J. grandiflorum* the flowers are held in a branched and forked inflorescence.

Both this species and *Jasminum polyanthum* are widely grown but can be easily distinguished by the clear 3-veined leaflets and short calyx lobes of *J. polyanthum*.

Jasminum grandiflorum L., *Sp. Pl.* ed. 2: 9 (1762). Type: 'Habitat in Malebaria'. (Herb. LINN. 17.2).

Syn. *Jasminum officinale* L. var. *grandiflorum* (L.) Stokes.

J. officinale f. *grandiflorum* (L.) Kobuski.

ILLUSTRATIONS. Redouté, *Choix des plus belles Fleurs* t. 70 (1827); Menninger, *Flowering Vines of the World,* p. 250, fig. 204 (1970); Graf, *Exotica*, edn. 12, 2: 1633 (1985); *The Plantsman* 10: 155 (1988); *Botanical Register* 2: 91 (1816); Herklots, *Flowering Tropical Climbers* fig. 202 p. 140 (1976); Phillips and Rix, *Conservatory and Indoor Plants* 2: 103 (1997).

Fig. 12. *Jasminum grandiflorum*.

DESCRIPTION. Evergreen, scrambling, glabrous, weak climber. *Branches* round or angled, green. *Leaves* opposite, dark green, leathery, pinnate usually with 7 leaflets but sometimes with 5–11 leaflets, to 12 cm long; petiole 0.5–4 cm long. *Leaflets* ovate to elliptic, with 1–2 pairs of veins; base cuneate or blunt; apex acute or acuminate; terminal leaflet 0.7–3.8 × 0.5–1.5 cm; lateral leaflets smaller. *Inflorescence* open, terminal or axillary, 5–50-flowered, pedicels to 3 cm, those of later flowers exceeding those of the first or central ones. *Bracts* linear, 5 mm long. *Flowers* white sometimes tinged with red outside, strongly fragrant, 2–3.5 cm across. *Calyx* tube 1.5–2 mm long, lobes linear, 5–10 mm. *Corolla* tube *c.*15–22 mm, lobes 5, broadly elliptic, sometimes with slightly wavy margins, 8–17 × 4–6 mm. *Berries* black, ellipsoid, 6–9 mm (rare in cultivation).
DISTRIBUTION. Southern Arabia and NE Africa, Himalaya from China to India and Pakistan.
HABITAT. Bushland; 500–1,500 m.
FLOWERING TIME. Mainly summer but also sporadically from late spring into autumn.

Two subspecies have been recognised by Green (1986). They may be distinguished by their leaf characters: those of subsp. *grandiflorum* have more but slightly smaller leaflets than those of subsp. *floribundum*.

1. Leaflets (5–)7–9(–11), terminal leaflet (1–) 1.5–3.5 (–4) cm . . . . subsp. ***grandiflorum***
1a. Leaflets 3–5(–7), terminal leaflet (1.2–)2.5–3(–5) cm . . . . subsp. ***floribundum***

2a. subsp. **grandiflorum**

Plants of *Jasminum grandiflorum* from the Himalayas belong to subsp. *grandiflorum*. It is widely grown in warm temperate and tropical regions as a source of oil of jasmine for perfume and is also the national flower of Pakistan.

The cultivar 'De Grasse' has slightly larger flowers.

2b. subsp. **floribundum** (R. Br. ex Fres.) P. S. Green

Syn. *J. floribundum* R. Br. Type: Ethiopia on the way from Halei to Temben. *Rüppell* (FR).

The native habitat of subsp. *floribundum* is Arabia and Ethiopia, Sudan, Somalia, Rwanda, Uganda, Kenya, Yemen and Oman, usually in upland evergreen bushland at up to around 2,000 m. This subspecies forms a low shrub, scrambler or slender twiggy bush with pinnate leaves usually of 5 elongate ovate leaflets, with obscure veins and the terminal leaflet is distinctly larger than the laterals. The inflorescence has fewer flowers, the flowers tend to be slightly larger, to 3 cm across. It is also widely cultivated and the leaves and flowers have long been known in indigenous medicine; the leaves are astringent in action. The whole plant is considered to be anthelmintic and diuretic. In northern Europe it is best grown under glass but in southern Europe will tolerate slight frost.

3. JASMINUM POLYANTHUM

Section Jasminum

This species was discovered in 1883 by Père Delavay in Tali, Yunnan, and collected in 1891; Franchet named it, appropriately enough for its profusion of flowers. It has probably been introduced on more than one occasion, but it is certainly recorded that when Major Lawrence

Johnston went to Yunnan, China with George Forrest in 1931, he brought it back to his garden at Serre de la Madone near Menton. They saw it growing in hedges in the hills and noted it in the gardens of the customs office at Tengyueh (Tengchong).

Jasminum polyanthum is on the borderline of hardiness, but it is possible to grow it in very sheltered gardens in south and south-west England, and particularly in town gardens in London. Of course, it is very popular in Mediterranean climates and it also grows well in California. In colder regions it is usually grown in a cool conservatory, but as it is a vigorous plant when given a free

2 cm

Fig. 13. ***Jasminum polyanthum***. Line drawing by Geoffrey Herklots.

root run it would not be suitable for a very small space. It is widely sold as a pot plant and grown in very large quantities, especially for the Dutch flower market: by dropping the temperature, flowering can be initiated and encouraged to occur at a predetermined time. The masses of fragrant flowers, even on small plants, make it a popular houseplant although in the hot dry atmosphere of central heating, the flowers may not last as long as they would outside. After the frosts are over, the plant can be put outside in a sunny position. It is very easy to propagate by cuttings. When grown outside, cooler nights will result in deeper pink colouration of the buds and the outside of the corolla but plants grown indoors are often pure white. To produce the maximum number of flowers, it is important that the shoots be in full sun; even in hot summer climates, shaded plants produce few flowers. This species flowers early and even in the mildest gardens of northern Europe the young growth and flowers may be affected by a slight late frost; it may also have a second, less prolific flowering in late summer. This species is also cultivated commercially for its aromatic oil. Its hardiness rating is H5, USDA zones 8–11.

Jasminum polyanthum is very similar to *J. officinale* but is less hardy, coming from habitats with a lower altitude and latitude, and can be recognised by its 3-veined leaflets and very short calyx lobes. It is also related to *J. dispermum,* but in flower *J. dispermum* has a more compact inflorescence because of its considerably shorter pedicels.

Jasminum polyanthum Franch., *Rev. Hort. [Paris]* 1891: 270, fig. 69 (1891). Type: cult. ex Delavay, Yunnan, China (P?).
Syn. *Jasminum blinii* H. Lév., *Repert. Spec. Nov. Regni Veg.* 13: 151 (1914).
 Jasminum dealfieldii H. Lév., *Cat. Pl. Yunnan*, 179 (1916) *in adnot.*

ILLUSTRATIONS. *Curtis's Botanical Magazine* 161: t. 9545 (1938); *Journal of the Royal Horticultural Society* 91: f. 208 (1966); *The Plantsman* 10: 155 (1988); Herklots, *Flowering Tropical Climbers*, fig. 204 p. 141; Brickell, *The RHS A–Z Encyclopedia of Garden Plants* 586 (2003); Phillips and Rix, *Conservatory and Indoor Plants* 2: 102 (1997); Ellison, *Cultivated Plants of the World*, p. 329 (1995).
DESCRIPTION. Twining evergreen, or semi-evergreen in cooler locations, glabrous climber. *Branches* round or angular, slender, green. *Leaves* opposite, somewhat leathery, pinnate with 5–7 leaflets to 15 cm long, but decreasing in size towards the tips of the branches; petiole 5–20 mm. *Leaflets* 3-veined, dark green above, sometimes with tufts of hairs in vein axils beneath; terminal leaflet longer, lanceolate to narrow lanceolate; 3.5–9 × 1–3 cm, base rounded to truncate; apex acute to long acuminate; lateral leaflets 2–5.5 × 1–2.5 cm, ovate, shortly stalked, base rounded to truncate or oblique, apex acute to acuminate. *Inflorescence* axillary or terminal forming a leafy panicle, 7–16 cm long, 5–25 or more flowered; pedicels 1–2 cm or more, very slender. *Bracts* 1.5–3 mm. *Flowers* rose-pink in bud, white often tinged red or pink outside, strongly fragrant, 12–20 mm diameter. *Calyx* tube 1–2 mm long, lobes linear, 1–2 mm. *Corolla* tube 2–2.5cm, lobes 5, ovate-elliptic, 8–12 × 4–5 mm. *Berries* black, sub-globose *c.*7–6 mm.
DISTRIBUTION. SW China; (Yunnan, Guizhou and southern Sichuan).
HABITAT. Woods and scrub; from 1,300–2,200 m or higher.
FLOWERING. Spring, but at almost any time in the pot plant trade.

Plate 5. *Jasminum polyanthum*. Painted by Lilian Snelling for *Curtis's Botanical Magazine*, t. 9545 (1938).

4. JASMINUM DISPERMUM

Section Jasminum

Jasminum dispermum is a scrambling plant with leaves with 3 or 5 leaflets on the same branch, and can be distinguished from *J. polyanthum* by its smaller white summer flowers with shorter more rounded corolla lobes. The name *dispermum*, two seeded, refers to the two seeds in the berry, a character of several species.

This species has two main areas of distribution in the wild in the Himalayas. Plants from the western ranges, from Pakistan, India, Bhutan and Nepal, have leaves usually of 3 to 5, or even 7, leaflets, the terminal one usually 7–9 × 3–4.5 cm. These may be referred to subsp. *dispermum*. Plants from the eastern regions, India, Burma, China and Thailand usually have only 1 to 3 larger leaflets with the terminal one 9–11 × 4–6 cm. These are referred to subsp. *forrestianum* P. S. Green. Plants with simple leaves have been described as distinct species, *J. simonsii* Sinha and *J. scalarinerve* Kobuski, and the suggestion has been made that they should be included in section *Unifoliolata*, but Green (1997) puts these species firmly in *J. dispermum*.

Jasminum dispermum may survive outdoors in the mildest and the most sheltered parts of Europe in sun or partial shade but cannot tolerate temperatures lower than around -5° C, so generally it needs cool greenhouse conditions. However, it has never been widely cultivated in western gardens despite the good fragrance of the very numerous white flowers. Its hardiness rating is G1, USDA zones 9–11.

Records of the Royal Horticultural Society note that this species was shown at a flower show in February 1937 when it received an Award of Merit. The exhibitor was A. E. Osmaston of Wisborough Green in Sussex who wrote *The Forest Flora of Kumaon*, India, and who introduced the plant from there. However it is also mentioned in *Paxton's Magazine* in 1848 as a climbing shrub from Nepal with white flowers, strongly scented especially in the evening. Presumably plants of the original collection were lost to cultivation sometime before 1937.

Fig. 14. *Jasminum dispermum*.

Plate 6. *Jasminum dispermum*. Painted by Lilian Snelling for *Curtis's Botanical Magazine*, t. 9567 (1939).

Jasminum dispermum Wall. in Roxb., *Fl. Ind*. 1: 99 (1820). Type: Himalaya, *Wallich* 2886 (K). Syn. *Jasminum forrestianum* Kobuski, *J. Arnold Arbor*. 20: 71 (1939).

ILLUSTRATIONS. *Curtis's Botanical Magazine* 162: 9567 (1939); *Paxton's Magazine of Botany* 15: 174 (1848); Herklots, *Flowering Tropical Climbers* fig. 201 p. 140 (1976); Ellison, *Cultivated Plants of the World*, p. 329 (1995).

DESCRIPTION. Evergreen climbing glabrous shrub. *Branches* 4-angled, with stiff short hairs at the nodes. *Leaves* opposite, dark green, usually with 5 leaflets but sometimes fewer, petiole 1–2 cm long. *Leaflets* leathery, lanceolate to ovate, margins sometimes undulate, with 3–5 pairs of veins although the lower pair are more prominent and with hairs in the vein axils; base rounded, truncate to subcordate; apex acute or acuminate; terminal leaflet ovate, 6–9 × 3–4 cm; lateral leaflets much smaller, to 2–4.5 × 0.5–1 cm. *Inflorescence* lax, axillary or terminal, 3 to 50-flowered; pedicels 3–7 mm. *Bracts* linear, to 3 mm. *Flowers* white, purplish-red outside and in bud, fragrant, 10–15 mm diameter. *Calyx* tube 2 mm long, lobes minute, triangular. *Corolla* tube 1–1.5 cm, lobes 5, ovate to obovate, 6–7 × 5.5 mm. *Berries* purple-black, ovoid, *c*.1 cm long.

DISTRIBUTION. Temperate Himalaya, from Bhutan, India, Kashmir, Nepal, W China, Thailand and Burma.

HABITAT. Montane forests, thickets and hedges; from about 1,400–2,800 m.

FLOWERING TIME. Late spring to summer.

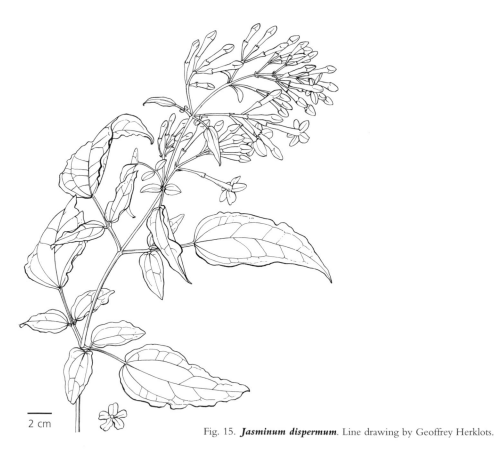

2 cm

Fig. 15. *Jasminum dispermum*. Line drawing by Geoffrey Herklots.

5. JASMINUM BEESIANUM

Section Jasminum

As *Jasminum beesianum* is a weakly twining climber, it is best grown with support, although it can be used as ground cover or allowed to scramble through other shrubs or trees. It is easy to grow and readily layers itself where the branches touch the ground. Although hardy, it will lose its leaves in winter in exposed or cold situations. It is noted for attractive and abundant glossy black berries which persist on the plant for a long period, especially if one or more clones are grown together. It is moderately frost-hardy, and is generally H3, USDA zones 8–11.

Jasminum beesianum can be recognised easily by its combination of red or reddish purple flowers and simple leaves.

With its simple leaves, it would be expected that *Jasminum beesianum* would be classified with the other species of the section *Unifoliolata*, but the existence of *J.* × *stephanense,* a natural and artificial hybrid with the pinnate leaved *J. officinale*, implies that it is more closely related to and should be classified in the section *Jasminum*. This is supported by the fact that the petioles are not articulated, a character of members of the section *Unifoliolata*, and also by certain anatomical seed characters such as the endosperm and layers of the seed coat (Rowher, 1996).

Jasminum beesianum was first collected by French missionaries around 1895, and then by E. H. Wilson in 1904. In *c.*1906 it was collected by George Forrest for the nursery firm Bees Ltd., who first offered it for sale in 1910 describing it in their catalogue as rampant, very hardy and with bright red flowers. The name commemorates the nursery that was one of Forrest's main sponsors.

Jasminum beesianum Forrest & Diels in Diels, *Notes Roy. Bot. Gard. Edinburgh* 5: 253 (1912). Type: China, Yunnan. Forrest 2021 (E).
Syn. *Jasminum beesianum* var. *ulotrichum* Hand.-Mazz.
    J. valbrayi H. Lév., *Repert. Spec. Nov. Regni Veg.* 13: 337 (1914).
    J. violascens Lingelsh., *Repert. Spec. Nov. Regni Veg. Beih.* 12: 463 (1922).
    J. delavayi Franch. ex Diels., *Notes Roy. Bot. Gard. Edinburgh* 5: 253 (1912).

ILLUSTRATIONS. *Gardeners' Chronicle* 77: 130 (1925) in fruit; *Curtis's Botanical Magazine* 151: 9097 (1926); *The Plantsman* 10: 155 (1988); Phillips and Rix, *Shrubs* p. 100 (1989).

DESCRIPTION. Deciduous glabrous or minutely hairy weak climber. *Branches* 4-angled, green. *Leaves* opposite, rather dull green, simple, broad-ovate to ovate-lanceolate, 1–4 × 0.3–1.8 cm, with 1–3 pairs of obscure veins; base rounded; apex long acute to long acuminate; petiole 1–2 mm long. *Inflorescence* sub-umbellate, terminal on short leafy side shoots; usually 3-flowered, pedicels 5–15 mm. *Bracts* linear, 4–10 mm. *Flowers* red, reddish purple or pink, fragrant, 7–15 mm diameter. *Calyx* glabrous or hairy, tube 2 mm long, lobes 5–7, linear, 3–10 mm. *Corolla* tube, 0.9–1.5 cm, hairy in throat, lobes usually 6, ovate-orbicular, 2–7 × 2–5 mm. *Berries* black, glossy, globose or ellipsoid, *c.*6 mm long.

DISTRIBUTION. South-western China, Yunnan, Guizhou, Sichuan, Xizang.

HABITAT. Slopes, grasslands, open rocky ground, thickets and woods, on the banks between paddy fields; 1,000–3,600 m.

FLOWERING TIME. Summer; May onwards in Yunnan.

6. JASMINUM ×STEPHANENSE

This hybrid between *Jasminum beesianum*, the pollen parent, and *J. officinale* (a large flowered form) the seed parent, and intermediate in all its characters, was raised at St Etienne in France by Thomas Javit. It is easily recognised as a hardy climber with scented pale pink flowers and compound leaves. It was shown at the Société Nationale d'Horticulture de France in July 1820 and first sold by the well-known French nursery Lemoine and Sons around 1921. The French missionary, Père Delavay first found this hybrid in the wild in 1887 and Joseph Rock collected it in Yunnan in 1922.

M. Javit tried crosses with other species but without success. He did however obtain one seed from *Jasminum ×stephanense* so the hybrid is not completely sterile. M. Léon Chenault had written in *Revue Horticole,* in 1913, that a good plant could be developed if more scent could be brought into the small but numerous red flowers of *J. beesisanum* by hybridisation, and this is exactly what is what Javit did seven years later. This is the only well-recognised hybrid in the genus both found wild and raised deliberately in cultivation, but there are attempts to produce new and exciting hybrids which have not yet come to fruition.

The flower colour tends to be brighter in cooler weather and the plant, though easy to grow, is best in a light soil against a sunny wall. It is somewhat hardier than *Jasminum beesianum* and is usually rated H4, USDA zones 7–11.

The name *stephanense* is derived from Stephanus, the Latin equivalent of St Etienne where the hybrid was first developed.

Jasminum ×stephanense Lemoine, *Cat.* No. 195, pl. 9 (1921). Type: Not traced.

ILLUSTRATIONS. *Revue Horticole* 20: 644 (1927); Gault, *Dictionary of Shrubs in Colour,* f.11/259 (1976); Phillips and Rix, *Shrubs* p. 233 (1989).

DESCRIPTION. Evergreen climber. *Branches* angled, pubescent, green. *Leaves* opposite, simple to pinnate or pinnatisect to 10–13 × 5–10 mm, pubescent when young; veins obscure; petiole 5–20 mm. *Leaflets* 1–9, but usually 3–5, ovate to elliptic, slightly hairy below; base rounded; apex acute to shortly acuminate; 5–40 × 2–15 mm. *Inflorescence* terminal on side shoots, solitary or 3–10-flowered, pedicels 5–20 mm. *Bracts* linear, 6–10 mm. *Flowers* pink or reddish especially on outside, fragrant, to 20 mm diameter. *Calyx* minutely hairy, tube 2–4 mm long, lobes linear, 2–4 mm. *Corolla* tube 10–18 mm, lobes 4–6, ovate to broadly elliptic, 6–9 × 3–4 mm long. *Berries* black, subglobose, 5–7 mm across.

DISTRIBUTION. SW China, in Sichuan, Xizang, Yunnan.

HABITAT. Thickets, woods, ravines; 2,200–3,100 m.

FLOWERING TIME. Summer.

Plate 7. *Jasminum beesianum.* Painted by Lilian Snelling for *Curtis's Botanical Magazine,* t. 9097 (1925).

Section TRIFOLIOLATA DC.

Section *Trifoliolata* is recognised by its white flowers and opposite, usually trifoliolate, leaves with petioles that are not articulated. The flowers are sometimes rather small; the fruit is usually one-seeded, except in *Jasminum abyssinicum* which is closer to Section *Jasminum* in seed characteristics. The species range from southern, eastern and tropical Africa, Madeira, India, the Himalayas, south-western China and Australia. The type species is *J. didymum* G. Forst.

7. JASMINUM ABYSSINICUM

Section Trifoliolata DC.

This species was described from Abyssinia, but is now known from much of East Africa, south to Tanzania and west to Congo (Kinshasa). In the wild, and with support in cultivation, it may reach 15 m or more. Herklots (1976) records how he saw it reaching over 15 m on the margins of forest near Bonga, and both wild and cultivated in other parts of Ethiopia where it flowers from January until March. Plants with both pure white and pink-flushed flowers were found in the same area. The flowers are rather small, but the species is very floriferous and the flowers have a strong and heavy scent.

Jasminum abyssinicum is similar to *J. fluminense* but it has a more open oval shaped, rather than flat-topped inflorescence and larger, non-leathery leaves. It has been cultivated for many years in the Royal Botanic Gardens at Kew where it thrives in the Temperate House, which is kept just frost-free during the winter. Its hardiness rating is G2, USDA zones 10–11.

Jasminum abyssinicum Hochst. ex DC., *Prodr.* 8: 311 (1844). Type: Aduwa, Abyssinia. *Schimper* 169 (isotype M).

ILLUSTRATIONS. Herklots, *Flowering Tropical Climbers*, fig. 197, p. 138 (1976); Kupicha, in *Flora Zambesiaca* 7 (1): tab. 70; Phillips and Rix, *Conservatory and Indoor Plants* 2: 105 (1997).

DESCRIPTION. Evergreen climber to 15 m glabrous or slightly hairy on bracts and calyx. *Branches* round. *Leaves* opposite, trifoliolate, petiole 1.5–3.3 cm long. *Leaflets* ovate to broad-ovate, with *c.*4 pairs of veins with tufts of stellate hairs in vein axils; base cuneate to rounded; apex acute; terminal leaflet variable but usually to 10 × 5 cm; lateral leaflets to 7.5 ×

Fig. 16. *Jasminum abyssinicum*.

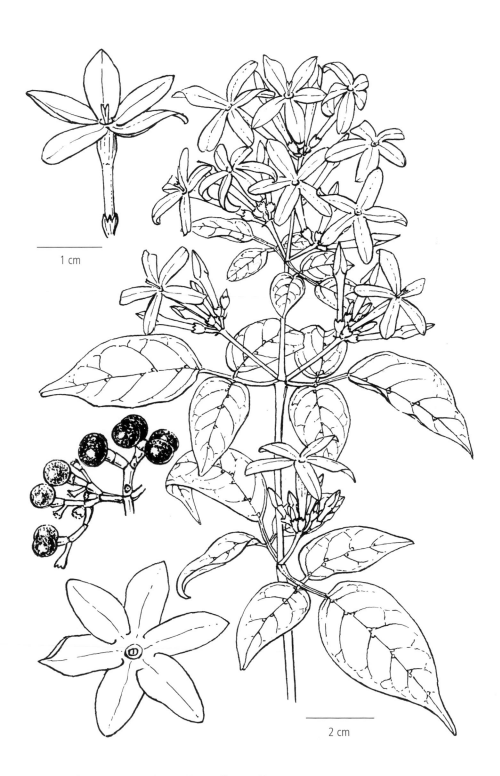

Fig. 17. *Jasminum abyssinicum*. Line drawing by Geoffrey Herklots.

4.4 cm. *Inflorescence* terminal on side shoots and axillary, 20–50-flowered, pedicels 4–15 mm. *Bracts* linear, 2–5 mm. *Flowers* white, sometimes flushed pink in bud and on outside of corolla, fragrant, to 2 cm across. *Calyx* hairy, tube 3–4.5 mm long, lobes minute. *Corolla* tube *c.*1.3–2.5 cm, lobes 4–6, oblong-elliptic, 7–10 × 4–6 mm. *Berries* black, globose, *c.*10 mm.

DISTRIBUTION. East Africa, from Ethiopia to Tanzania and Congo (Kinshasa).

HABITAT. Streams and forest edges or undergrowth, mostly in evergreen montane forest; 650–2,830 m.

FLOWERING TIME. Winter to early spring.

8. JASMINUM ANGULARE

Section Trifoliolata DC.

This is a variable species in the wild, and as it has been brought into cultivation on more than one occasion there may be several variants in gardens. It was grown by Jacquin (1804) and illustrated in his book on plants in the gardens of the Imperial Palace of Schönbrunn near Vienna at end of the eighteenth century. A Mrs Birks first brought it to Britain from the Cape and it first flowered at the Royal Botanic Gardens, Kew in 1885 when it was painted for the account in *Curtis's Botanical Magazine*.

Variations in pubescence appear to bear no relation to geographical location, nor to the recognition of varieties. For example, var. *glabratum* E. Mey. is dismissed by Verdoon (1956). Even the most hairy-leaved forms, although similar in the long corolla tube to *Jasminum sinense*, can be distinguished by their shorter calyx lobes. Another characteristic of *J. angulare* is the manner in which the petioles are held rather erect so that the branches appear denser and the upper leaves tend to hide the base of the inflorescence.

In its native country *Jasminum angulare* will survive a mild frost, is drought resistant and is used to climb over trellis or fences and to cover banks. It is used in local traditional medicine and is said to protect against lightning. The fresh leaves however are considered poisonous to cattle and sheep.

Jasminum angulare can be recognised by its flowers which have an exceptionally long, narrow corolla tube, and the trifoliate leaves which are never very hairy like those of *J. sinense*. This species is named *angulare* because of its angled branches.

Fig. 18. *Jasminum angulare*.

Plate 8. *Jasminum angulare*. Hand-coloured lithograph by Matilda Smith for *Curtis's Botanical Magazine*, t. 6865 (1886).

Jasminum angulare Vahl, *Symb. Bot.* 3: 1 (1794). Type: habitat ad Cap. b. Spei, *Drege* s.n. (C). Syn. *J. capense* Thunb., *Prodr. Fl. Cap.* 2, (1794).

ILLUSTRATIONS. *Curtis's Botanical Magazine* 112: 6865 (1886); Herklots, *Flowering Tropical Climbers*, fig. 198, p. 138 (1976); Brickell, *The RHS A–Z Encyclopedia of Garden Plants* 585 (2003); Phillips & Rix, *Conservatory and Indoor Plants* 2: 105 (1997).

DESCRIPTION. Evergreen glabrous or hairy climber or straggly shrub to 1.5 m or more. *Branches* ribbed, green. *Leaves* opposite, hairy or glabrous, usually trifoliolate, occasionally some leaves with 5 leaflets, dark green, glossy; petiole 3–20 mm long. *Leaflets* ovate to broad-ovate to oblong, with 3–4 pairs of veins with hairs in axils especially of lower pairs of veins; base obtuse to rounded; apex acute to rounded, mucronate; terminal leaflet 1.3–4.5 × 0.6–2.5 cm; lateral leaflets smaller, to 1–2 × 0.5–1 cm. *Inflorescence* compact, terminal or axillary, usually 3–5 flowered; pedicels 1–2 cm. *Flowers* white, slightly greenish or pink tinges on outside, fragrant, 2–3.5 cm across. *Calyx* tube 2.5–3 mm long, lobes *c*.1–2 mm, triangular. *Corolla* tube 1.7–3.5 cm, lobes 5–6, oblong to ovate, 10–15 × 6–7 mm. *Berries* bluish black, globose, *c*.7 mm.

Fig. 19. *Jasminum angulare.*

DISTRIBUTION. Widespread in South Africa from the Eastern Cape to the Transvaal.
HABITAT. Coastal bush, hillsides and in scrubland.
FLOWERING TIME. Summer.
Cultivars:

'Anne Shelton'

An almost hardy selection to 2 m tall, with white star-shaped flowers, large for the species, produced in summer and then again in autumn and early winter if grown under glass. In sun or partial shade, it may survive outdoors in the milder parts of Europe so long as there is no more than very slight frost. Its hardiness rating is G1, USDA zones 10–11.

2 cm

Fig. 20 (right). *Jasminum angulare.*
Line drawing by Geoffrey Herklots.

9. JASMINUM AURICULATUM

Section Trifoliolata DC.

Jasminum auriculatum has been known in England at least since 1818, when it was described and illustrated in *Edwards's Botanical Register*; there it is said to have been introduced by Sir Joseph Banks in 1790. It is now rarely seen in western gardens but is grown in India as a source of jasmine oil. It is also cultivated in Thailand and Pakistan. The flowers are slightly smaller than many species and perhaps not such a pure white, but amply compensated for by their especially sweet scent and thick texture. The leaves are quite variable in hairiness, with some forms being densely felted and silvery whereas others are almost glabrous.

It can be recognised by its trifoliolate leaves with very small lateral leaflets; when the lateral leaflets are absent, the petiole is jointed as it is in the majority of the species in Section *Unifoliolata* with simple leaves.

Jasminum auriculatum is said to grow well in sunny positions in containers in European gardens, making a compact, bushy plant to about 2 m tall with tight heads of flowers. It probably requires frost-free conditions. Its hardiness rating is G2, USDA zone 11.

The name *auriculatum,* meaning with auricles or small ears, refers to the very small lateral leaflets.

Jasminum auriculatum Vahl, *Symb. Bot.* 3: 1 (1794). Type: colitur in hortis idolatrorum Malabariae, *König*, or Isle de Bourbon, *Commerson*, Herb. Vahl (C).

ILLUSTRATION. *Botanical Register* 4: 264 (1818).

DESCRIPTION. Evergreen softly hairy, or almost glabrous climber. *Branches* tomentose. *Leaves* opposite, trifoliolate, sometimes simple, grey pubescent; petiole 1–2.5 mm long. *Leaflets* ovate, with 2–3 obscure veins; base rounded; apex acute, mucronate; terminal leaflet to 2–5 × 0.8–2 cm; lateral leaflets much smaller, rarely over 4 mm across. *Inflorescence* terminal on side shoots, 5- to many-flowered, pedicels to 5 mm. *Bracts* linear, 4 mm long. *Flowers* white, fragrant, 1.8–2 cm. *Calyx* hairy, tube 1.5 mm long, lobes minute. *Corolla* tube 1.5 cm, lobes 5–7, elliptic, up to 8 mm long. *Berries* black, globose, 5 mm.

DISTRIBUTION. India, NW Himalayas and Nepal to Sri Lanka, mainly in rather dry regions.

HABITAT. Thickets, ravines, scrubland and forests; 800–4,500 m.

FLOWERING TIME. Summer to autumn in monsoon climates. July to December in Sri Lanka.

10. JASMINUM AZORICUM

Section Trifoliolata DC.

Despite its name, *Jasminum azoricum* is native to Madeira, not the Azores, although it is cultivated in gardens there. It has also been cultivated in many other countries around the world and has become naturalised in parts of Brazil and the Caribbean, although some of these records may be of *J. fluminense* with which it has been confused. *Jasminum fluminense* differs in its shorter pedicels, rarely more than 3 mm, a slightly shorter corolla tube and more hairs, which are crisped, on the inflorescence. *Jasminum azoricum* is also close to *J. angulare* but is a more vigorous plant with lanceolate and more pointed corolla lobes, giving the flower a very star-like appearance.

This species was brought into cultivation in Europe towards the end of the seventeenth century and is not difficult to grow with sufficient heat. It has a woody base from which it puts out long shoots in summer, and will flower in both sun and partial shade for a long period from spring to autumn. If planted in a warm position, it can survive a few degrees of frost, and in England it will overwinter in an unheated greenhouse. Its hardiness rating is G2–G1, USDA zones 9–11.

Fig. 21. *Jasminum azoricum*.

Plate 9. *Jasminum auriculatum*. Hand-coloured engraving by Sydenham Edwards from *Botanical Register* 4, t. 264 (1818).

Fig. 22. *Jasminum azoricum*.

Green (1965) notes that there was a record in 1770 of a form with gold variegated leaves, but no specimens have been seen.

Jasminum azoricum L., *Sp. Pl.* 7 (1753). Type: Herb. Linn. 1714 (LINN).
Syn. *J. trifoliatum* Moench., *Meth. Pl.* 467 (1794).

ILLUSTRATIONS. *Botanical Register* 1: t. 1 (1816); *Curtis's Botanical Magazine* 44: 1889 (1817); Phillips & Rix, *Conservatory and Indoor Plants* 2: 105 (1997).

DESCRIPTION. Evergreen glabrous scrambling and twining climber to 7 m. *Branches* rounded. *Leaves* opposite, trifoliolate, glossy, somewhat leathery; petiole 1–2.5 cm long. *Leaflets* ovate to ovate-lanceolate, often folded along midrib, with 5–6 pairs of veins; base obtuse to rounded; apex acute to shortly acuminate; terminal leaflet to 3–9 × 2–5 cm; lateral leaflets slightly smaller. *Inflorescence* an open panicle of terminal and axillary cymes, 3–20 cm, 5–25-flowered, pedicels to 10 mm. *Bracts* linear to 3 mm long. *Flowers* white, sometimes purple flushed in bud, fragrant, size 2–2.5 cm across. *Calyx* glabrous, tube 2–3 mm long, lobes triangular, *c*.1 mm. *Corolla* tube 15–20 mm, lobes 4–6, oblong-lanceolate, acute, 10–15 × 5 mm. *Berries* not seen.

DISTRIBUTION. Endemic to Madeira.

HABITAT. On cliffs and in ravines, but rare and localised; at 900–1,000 m or more.

FLOWERING TIME. Almost any time from late winter to autumn depending on conditions.

Plate 10. *Jasminum azoricum*. Hand-coloured engraving by an unknown artist from *Curtis's Botanical Magazine*, t. 1889 (1817).

11. JASMINUM DIDYMUM

Section Trifoliolata DC.

Jasminum didymum has a wide geographical distribution across the Pacific and is also very variable. It can be recognised by its trifoliate leaves with numerous small flowers borne in a large paniculate inflorescence. The name *didymum*, twinned, refers to the 2-lobed fruit. It has been separated into 3 subspecies by Green (2001). All three subspecies are cultivated in Australia, although subsp. *didymum* is the one that is usually grown in gardens of Europe and North America. This was introduced to Kew in the mid-nineteenth century, possibly by William Grant Milne (d. 1866), a collector on the surveying voyage of Captain Denham in the Pacific around 1855.

In Australia this subspecies is commonly known as the coastal jasmine, and is characteristic of stabilised sand dunes on the edge of coastal forests. It is useful for gardens in coastal conditions in an open position. Regular pruning will ensure it remains bushy.

Jasminum didymum G. Forst., *Fl. Ins. Austr. Prodr.* 3 (1786). Type: Society Islands (Tahiti) *Forster* (isolectotype BM, lectotype K).

ILLUSTRATION. *Curtis's Botanical Magazine* 104: 6349 (1878).
DESCRIPTION. Evergreen, slightly hairy or glabrous scrambling shrub climber to 6 m. *Branches* angled. *Leaves* opposite, usually trifoliolate, bright green and glossy; petiole 3–35 mm long. *Leaflets* narrow ovate to elliptic to linear, margins often recurved, with 3–6 pairs of veins sometimes with tufts of hairs in the axils of the veins; base subcordate to acute to rounded; apex acute to obtuse, apiculate; terminal leaflet 0.8–10 × 0.4–7.5 cm; lateral leaflets 0.5–9 × 0.3–6 cm. *Inflorescence* terminal on side shoots or axillary, 2–30 cm, 5–50 or more-flowered, pedicels 1–10 mm. *Bracts* triangular to linear, 0.5–2 mm long. *Flowers* white to cream, very fragrant, 1 cm or less across. *Calyx* tube 1.5–2 mm long, lobes 4–6, to 1 mm. *Corolla* tube to 4–10 mm, lobes 4–7, ovate-lanceolate, acute to obtuse 2–5 mm long. *Berries* purplish-black, globose *c*.8–12 mm.
DISTRIBUTION. Indonesia and Philippines to Tahiti, New Caledonia and northern Australia.
HABITAT. Open areas of rainforest or forest especially in coastal areas often on sand dunes or coralline rock.
FLOWERING TIME. Winter.

The three subspecies may be separated as follows:

1. Terminal leaflets at least 4 times as long as wide, rarely over 1 cm wide . . . . . . subsp. *lineare*
1a. Terminal leaflets less than 4 times as long as wide **2**
2. Terminal leaflet more than 4 cm long, more or less leathery, inflorescence 15–50-flowered . .
    . . . . subsp. *didymum*
2a. Terminal leaflet less than 4 cm long, not leathery, inflorescence 9–27-flowered . . . .
    . . . . subsp. *racemosum*

Plate 11. *Jasminum didymum* subsp. *didymum*. Hand-coloured lithograph by Harriet Thistleton-Dyer from *Curtis's Botanical Magazine*, t. 6349 (1878).

11a. subsp. **didymum**

This subspecies has a thick terminal leaflet, usually ovate to broadly ovate but sometimes narrower, 3.5–25 × 1–7.5 cm. The inflorescence is of up to 50 or more flowers, with stout pedicels; each flower is up to 10 mm across with 5 or 6 ovate corolla lobes, 3–6 × 3–4 mm.

11b. subsp. **lineare** (R. Br.) P. S. Geeen. Type: South Australia *Brown* s.n. holotype of *J. lineare* R. Br. (BM).
Syn. *J. lineare* R. Br., *Prodr.*: 521 (1810).

This subspecies grows wild in many parts of Australia in inland areas, in woodland especially on sandy soil. The common name desert jasmine indicates that it is useful for hot dry regions but with fewer flowers in each inflorescence it is less striking. This is a more straggly and prostrate plant with very narrow, dull green leaflets; the terminal leaflet is linear, 2.7–12 × 0.2–1.2 cm. The inflorescence is of up to 15 flowers with slender, but often very short, pedicels and each flower is 7–10 mm across with 5 or 6 ovate corolla lobes, 3–6 mm long.

Fig. 23. *Jasminum didymum* subsp. *lineare*.

11c. subsp. **racemosum** (F. Muell.) P. S. Green. Type: Queensland, 'Morton Bay' *Hill & Mueller* s.n. holotype of *J. racemosum* F. Muell. (K).
Syn. *J. racemosum* F. Muell., *Fragm*. 1: 19 (1858).

This subspecies is found in Queensland, in dry scrub forest and mixed woodland and rainforest at up to 460 m and in sandy areas near the sea. It is a more delicate and less showy plant than the other subspecies, with smaller leaves. It forms a climbing or creeping shrub usually less than 1.5 m tall. The terminal leaflet is less thick, narrowly lanceolate to ovate or orbicular, 0.8–5.5 × 0.4–2 cm. The inflorescence is of up to 25 or more flowers with slender pedicels, and each flower is 4–9 mm across with 5 or 6 narrow corolla lobes, 3–4 mm long.

The Australian species *Jasminum dallachii* F. Muell., Fragm. 4: 150 (1864), was once included as a variety of *Jasminum didymum* as var. *dallachii* (F. Muell.) Domin, named after J. Dallachy who first collected the plant in Queensland, near Rockingham Bay. It is found from SE Queensland to New South Wales as a climber of rain forests and on forest margins, from 700–1,000 m. The main difference from *J. didymum* is that the plant is pubescent-tomentose, hence its common name, soft jasmine, and there are obvious tufts of hairs in the axils of the veins of the leaflets. As far as is known this species is not in cultivation.

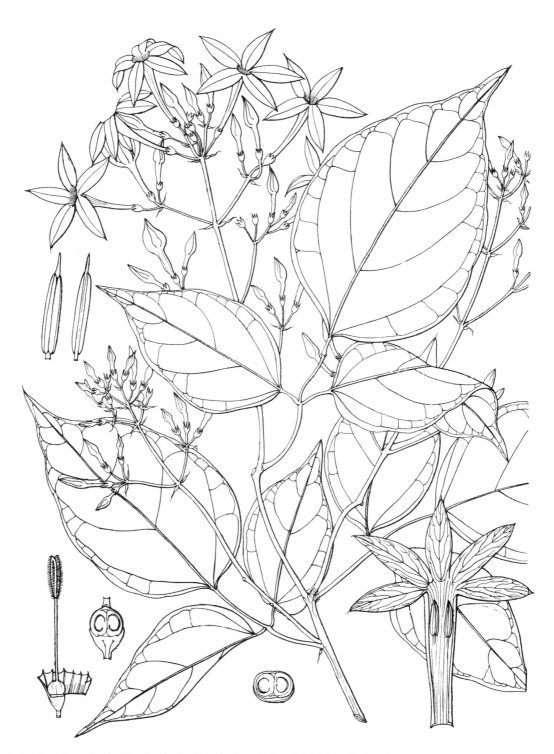

Fig. 24. *Jasminum flexile*. Line drawing by Rungiah from Robert Wight, *Icon. Pl. Ind. Orient.* 4, t. 1254 (1848).

12. JASMINUM FLEXILE

Section Trifoliolata DC.

Jasminum flexile is a climber with trifoliolate leaves, and can be recognised by having the leaflets with a submarginal vein and numerous small tufts of hairs in the vein axils. The flowers are white, with 5 oblong lobes and a tube 10–25 mm long.

It is widespread in SE Asia and is thought to have been grown in western gardens since the middle of the nineteenth century although there is no definite record of its initial introduction and it is seldom mentioned in the horticultural literature. The specimen grown at the Royal Botanic Garden, Edinburgh was collected by Kingdon Ward in Burma (Myanmar), in the Chin hills, probably during his last Himalayan expedition in 1956. It can tolerate only a few degrees of overnight frost, so its hardiness rating is G2, USDA zones 10–11.

Var. *hookerianum* Clarke from the eastern Himalaya in Khasia and Assam, differs in having membranous, broader leaflets to 6 cm wide, and larger flowers with broader lobes.

Jasminium flexile is not dissimilar in flower to *J. lanceolaria* but quite distinct in leaf characters and is distinguished from *J. fluminense* by its larger flowers but very small calyx lobes.

The name *flexile* refers to the slender flexible branches.

Jasminum flexile Vahl, *Symb. Bot.* 3: 1 (1794). Type: Habitat in India orientali, *König* (C).

ILLUSTRATION. Wight, *Icon. Pl. Ind. Orient.* 4 t. 1253 (1848).
DESCRIPTION. Evergreen woody climber. *Branches* round. *Leaves* opposite, trifoliolate; petiole 3–4 cm long. *Leaflets* ovate-lanceolate, membranous, with 3–5 pairs of veins with small tufts of hairs in the axils of the veins and the basal pair of veins forming a submarginal vein; base cuneate, rounded to truncate; apex acute to acuminate, terminal leaflet 7–10 × 3.5–5 cm; lateral leaflets 6–8.5 × 3–4 cm. *Inflorescence* terminal on side shoots or axillary, 7–11 cm long, 7–15 flowered, pedicels 3–7 mm. *Bracts* linear, 1.5–2 mm. *Flowers* white, fragrant, 2–3 cm diameter. *Calyx* tube 2 mm, lobes minute. *Corolla* white; tube 10–25 mm, lobes 5, oblong, 15 × 4 mm. *Berries* black, ellipsoid, *c*.8 × 6 mm.
DISTRIBUTION. SW China, India, (with var. *hookerianum* Clarke in Khasia and Assam), Sri Lanka, northern Thailand and Burma.
HABITAT. Evergreen forests, usually in rather moist and intermediate regions; to *c*.1,300 m.
FLOWERING TIME. July to December (in Sri Lanka).

13. JASMINUM FLUMINENSE

Section Trifoliolata DC.

Though a native of Africa, *Jasminum fluminense* has long been naturalised in several parts of tropical South America and the Caribbean. It was introduced to Brazil by the Portuguese in the early part of the nineteenth century, and it was from naturalised plants in Brazil that it was first described. The species is widely grown in tropical gardens in Central and South America and Hawai'i, where it is easy to grow and will tolerate hot and dry conditions. It is probably not grown in Europe, even under glass. Its hardiness rating is G2, USDA zone 11.

In its native country, it is widespread through tropical Africa. As it shows considerable variation, in African floras it is separated into a number of subspecies based on leaflet size and shape. Herklots (1976) records seeing it at 1,300 m between Dessie and Tendaho in Ethiopia, flowering in early February.

It has been eaten as a vegetable in parts of its native country and the trailing stems are used for making rope.

Jasminum fluminense is a climber or scrambler, with trifoliolate leaves and broad, non-leathery leaflets; the rather small but very fragrant flowers have short flower stalks. Though it is often confused with *J. azoricum*, the shorter flower-stalks of *J. fluminense* produce flowers in tighter heads and the first flower in the inflorescence is at a lower level than the later ones. It is also similar to *J. angulare* but has a smaller calyx and broader inflorescence.

The name *fluminense* refers to the river Amazon at Rio de Janeiro, from where it was first described.

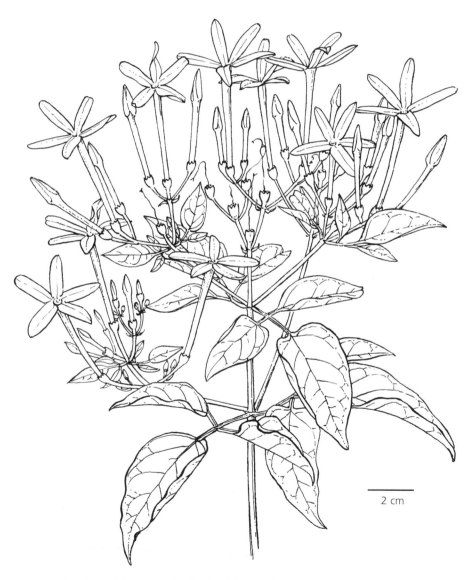

2 cm

Fig. 25. ***Jasminum fluminense***. Line drawing by Geoffrey Herklots, Dubte, Ethiopia, 5 Feb. 1972.

Jasminum fluminense Vell., *Fl. Flumin.* 10; i. T. 23 (1825). Type: Brazil, Santa Cruz 1825.
Syn. *J. bahiense* DC., *Prodr.* 8: 311 (1844).
   J. mauritianum Bojer ex DC., *Prodr.* 8: 310 (1844).

ILLUSTRATION. Herklots, *Flowering Tropical Climbers*, fig. 199, p. 139 (1976).
DESCRIPTION. Evergreen conspicuously hairy to glabrous weak climber or scrambling shrub. *Branches* round. *Leaves* opposite, trifoliolate, petiole 2–20 mm. *Leaflets* ovate to broadly ovate, with 4 pairs of obscure veins, often with hairs in the axils of the lower veins; very variable in size *c.*2–5 × 1.5–3.5 cm, base rounded to subcordate; apex acute to acuminate. *Inflorescence* on short leafless branches, in compact clusters, 5–100 or more-flowered; pedicels 1.5–6 mm. *Bracts* linear, 2–3 mm. *Flowers* white, buds slightly pink, fragrant, to 2–2.5 cm diameter. *Calyx* hairy, tube 1.5–3 mm long, lobes blunt, 1–1.5 mm. *Corolla* tube pale green to yellowish-white, 10–35 mm, lobes 6–7, rounded-elliptic, acute, 7–10 × 5 mm. *Berries* brownish-black, globose, *c.*7 mm.
DISTRIBUTION. Tropical Africa, Arabia and the Seychelles, naturalised in many parts of the New World tropics.
HABITAT. Forest, woodlands and grassland, especially along river courses; 210–1,950 m.
FLOWERING TIME. In flower and fruit most of the year.

14. JASMINUM LANCEOLARIA

Section Trifoliolata DC.

This is a robust climber with trifoliolate leaves, very leathery leaflets and flowers in dense clusters with very short pedicels. Herklots (1976) records that it is the commonest species in Hong Kong, and it is recorded as far west as Mount Omei and Ya-an in western Sichuan by Rehder (1916), when describing the collections made by E. H. Wilson and Augustine Henry. Hemsley (1889) described var. *puberulum* from Henry's collections in the Ichang gorges in Hubei.

   Jasminum lanceolaria is known to have been cultivated in the Botanic Gardens in Calcutta from plants collected from Silhet, but it is now not widely grown, if at all, in western gardens.

   The specific name refers to the narrow, lance-shaped leaflets. It should be noted that Roxburgh's spelling appears to be intentional and therefore should remain 'lanceolaria' and not altered to 'lanceolarium' (Green, 1995). It is probably frost-tender, hardiness rating is G 1–2, surviving in USDA zones 10–11.

Jasminum lanceolaria Roxb., *Fl. Ind.* 1: 97, 111 (1820). Type: Khasia, Moosman in sylvis, *Griffiths* in herb East India Co. 3715 (K, neotype, selected by P. S. Green).
Syn. *J. paniculatum* Roxb., *Hort. Bengal.*: 3 (1814).
   J. attenuatum Roxb. ex G. Don, *Gen. Syst.* 4: 62
   (1857).

ILLUSTRATION. Herklots, *Flowering Tropical Climbers*, fig. 200, p. 139 (1976).
DESCRIPTION. Evergreen woody, glabrous climber. *Leaves* opposite, trifoliolate, petiole 1–2.5 cm long. *Leaflets* smooth, glossy, leathery, lanceolate to elliptic-oblong, occasionally broader, with 4–6 pairs of obscure veins; base obtuse to rounded; apex rounded to shortly acuminate; terminal leaflet 7–13 × 3.5–5 cm; lateral leaflets slightly smaller, 3–8 × 3–4 cm. *Inflorescence* terminal on side shoots or axillary,

6–13 cm long, 8- to many-flowered, pedicels 0–5 mm. *Flowers* white, sweetly fragrant, to 3.2 cm across. *Bracts* linear, 0.5–2 mm. *Calyx* tube 2 mm long, lobes minute. *Corolla* tube 18–25 mm, lobes 5–6, elliptic to oblong, acute, 12–15 × 4–5 mm. *Berries* black, glossy, globose, *c*.10 mm.

DISTRIBUTION. Bhutan, Assam, Burma, China (except the north), Thailand.

HABITAT. Evergreen forests, wooded ravines to *c*.1,600 m.

FLOWERING TIME. Spring to summer.

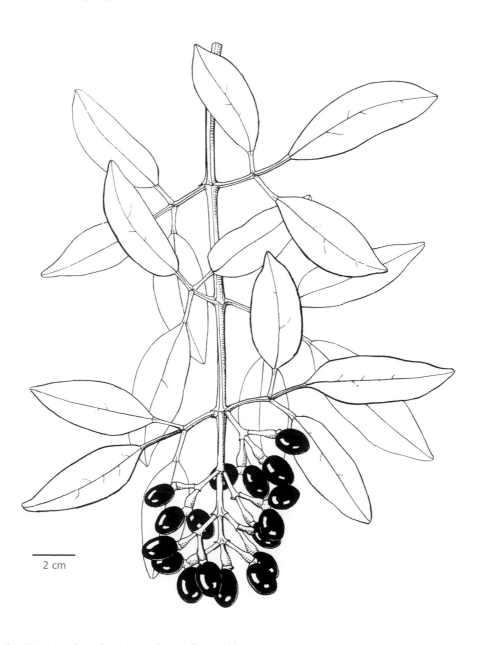

2 cm

Fig. 26. *Jasminum lanceolaria*. Drawn by Geoffrey Herklots.

15. JASMINUM SINENSE

Section Trifoliolata DC.

The date of the first introduction of this species into cultivation is uncertain but herbarium specimens of plants growing at Royal Botanic Gardens, Kew were taken as early as 1904, and P. S. Green in his account for *Curtis's Botanical Magazine* suggests that it came from a collection of 'C. Ford in August 1887 on a tour up the North and Lienchau Rivers in Kwangtung Province, China'. It was also collected by E. H. Wilson near Ya-an in western Sichuan at 600–1,000 m.

This attractive species is distinct and easily recognised by its very long slender corolla tubes, narrow calyx lobes which spread widely as the flowers droop and begin to fade, and soft velvety leaves which are pale grey-green underneath. The long corolla lobes resemble *Jasminum angulare,* but *J. sinense* has much larger ovate leaves. Unfortunately it is very tender and to grow anywhere other than the tropics it needs to be in a heated greenhouse, with the temperature kept around 10°C. Its hardiness rating is G2, USDA zone 10–11, though plants from Sichuan might prove hardier.

Jasminum sinense Hemsl., *J. Linn. Soc., Bot.* 26: 80 (1890). Type: syntypes: China: Hupeh [Hubei], Nanto and mountains to the northward, *A. Henry* 2106 and 4464 (K); Kwantung [Guangdong], North river, [August 1887], *Ford* 114 (K).
Syn. *J. bodinieri* H. Lév., *Fl. Kouy-Tcheou*: 294 (1914).

ILLUSTRATION. *Curtis's Botanical Magazine (Kew Magazine)* 13: t. 224 (1993).
DESCRIPTION. Evergreen hairy climber to *c.*8 m. *Branches* rounded, densely and softly hairy. *Leaves* opposite, trifoliate, densely tomentose, yellowish-green; petiole 1–2.5 cm long. *Leaflets* ovate to broadly ovate, with 4–5 pairs of veins, tomentose above, densely tomentose below; base obtuse to rounded; apex acute to acuminate; terminal leaflet 4–14 × 2.5–8 cm; lateral leaflets about half the size of the terminal leaflet. *Inflorescence* a usually dense terminal, 5–40-flowered, pedicels 0–5 mm. *Bracts* linear, 1–2 mm. *Flowers* white to yellow tinged, fragrant, 1–1.2 cm diameter *Calyx* hairy, tube 2–3 mm long, lobes linear to triangular, 1.5–3 mm long. *Corolla* tube pale green, 2.5–4 cm, lobes 5–6, oblong-elliptic, becoming slightly reflexed, 8–10 × 4 mm. *Berries* black, subglobose *c.*8 mm.
DISTRIBUTION. SW China (except Hainan) and Hubei.
HABITAT. Slopes, thickets and woods; below 2,000 m.
FLOWERING TIME. Intermittently through the year.

Plate 12. *Jasminum sinense.* Painting by Christabel King for *Curtis's Botanical Magazine,* t. 224 (1993).

16. JASMINUM TORTUOSUM

Section Trifoliolata DC

This South African species is not dissimilar to *Jasminum angulare*, *J. azoricum* and *J. fluminense* but has narrower leaflets not more than twice as long as broad and flowers with pedicels exceeding 10 mm. Its origin in European gardens is uncertain but it was known to be growing in Berlin when it was described by Willldenow in 1809. Jacquin also records it (as *J. flexile*) in the gardens of the palace of Schönbrunn, Vienna in 1804. It might survive outdoors in warm mild conditions. Its hardiness rating is G1, USDA zones 9–10.

The specific name, *tortuosum*, refers to the twisted or convoluted stems.

Jasminum tortuosum Willd., *Enum. Pl. Hort. Berol.* 1: 10 (1809). Type: from Cape Province, herb. *Willdenow* (B).
Syn. *J. campanulatum* Link, *Jahrb.* 1, 3: 30 (1820).
J. flexile Jacq., *Pl. Rar. Hort. Schoenbr.* 4: 46, t. 490 (1804).

ILLUSTRATION. Phillips and Rix, *Conservatory and Indoor Plants* 2: 105 (1997).

DESCRIPTION. Evergreen climber or scrambler. *Branches* angled. *Leaves* opposite, trifoliolate; petiole 1.5–2.5 cm long, sometimes to 20 mm, hairy. *Leaflets* linear to lanceolate, glabrous, with prominent midrib but obscure lateral veins, base cuneate; apex acute to rounded, mucronate; terminal leaflet 1.7–5 × 0.5–2 cm; lateral leaflets 2.5–3 × 1–1.5 cm. *Inflorescence* terminal and axillary, 3–5 flowered; pedicels 10–20 mm. *Bracts* linear, to 2 mm. *Flowers* white, fragrant, 30–35 mm diameter. *Calyx*, tube 2–3 mm long, lobes triangular, 1.5–2.5 mm. *Corolla*; tube 1.5–2.7 cm, lobes 5–6 oblong-elliptic, 12–20 × 4–5 mm. *Berries* not seen.

DISTRIBUTION. South Africa: in the Western Cape from Caledon to Mossel Bay.

HABITAT. Forest margins; 400–500 m.

FLOWERING TIME. Summer (November and December).

Fig. 27. *Jasminum tortuosum*.

Plate 13. *Jasminum tortuosum*. Hand-coloured engraving from Jacquin, *Hort. Schoenb*. 4, t. 490 (1804).

17. JASMINUM UROPHYLLUM

Section Trifoliolata DC

Jasminum urophyllum can be recognised by its leaves, which usually have three leaflets, though sometimes only a single leaflet; all are very clearly 3-veined, and the larger leaflets are distinctly acuminate. It was discovered by the Rev. Dr. Ernst Faber (1839–1899) of the Rhenish Missionary Society, who made one of the earliest collections of plants from Mount Omei in 1887.

Rehder described two new varieties from later collections in western China, in addition to a collection of var. *urophyllum* by Wilson from near Wa-shan, but they have been considered insignificant by later authors. Variants with narrower and totally glabrous leaves were named var. *wilsonii* Rehder, which was collected in western Sichuan, but without precise locality, under the number *Wilson* 4075. Var. *henryi* Rehder was described as having glabrous, usually simple leaves, oblong-lanceolate, rounded or truncate at the base, and lax cymes with slender pedicels 1–1.5 cm long.

The species is close to *Jasminum dispermum* but that is included in section *Jasminum* as the leaves usually have five leaflets.

The first record of this species' introduction into Britain appears to be seed sent to Royal Botanic Garden, Edinburgh from Nanking in 1936. It is, however, still very rare in cultivation and plants do not fruit as they are probably of only one clone. It does not appear in the Royal Botanic Garden, Edinburgh *Catalogue of Plants 2001*. It is generally considered to need greenhouse treatment, but from its habitat should survive in the open in areas with only light frost. Its hardiness rating is G2, USDA zones 10–11.

The name *urophyllum* refers to the elongated tip of the leaf, from the Greek, oura, meaning a tail.

Jasminum urophyllum Hemsl., *J. Linn. Soc., Bot.* 26: 81 (1889). Type: Szechuan, Mt Omei, 1523 m, *Ernst Faber* 47 (NY).
Syn. *J. urophyllum* var. *wilsonii* Rehder, *Plantae Wilsonianae* 2: 613 (1916).
  J. urophyllum var. *henryi* Rehder, *l.c.*

ILLUSTRATION. *Curtis's Botanical Magazine* 168: n.s. 148 (1951) as *J. urophyllum* var. *wilsonii*.
DESCRIPTION. Evergreen, glabrous or hairy climber to 2–3 m. *Branches* slightly angled, greenish brown. *Leaves* opposite, trifoliolate or simple at base of inflorescence, leathery, dark green, glabrous or hairy below ovate to elliptic to lanceolate, 3-veined; petiole 1–2 cm long. Simple leaves 5–17 × 2–5 cm; with 3 veins; base rounded; to subcordate; apex long acuminate. *Leaflets* of trifoliolate leaves lanceolate to ovate lanceolate, 3-veined; terminal leaflet larger, 4–9.5 × 1.5–3 cm; lateral leaflets 1–5 × 0.5–2 cm. *Inflorescence* open, glabrous to densely hairy, terminal or axillary, 8 cm across, 9–13-flowered, pedicels 1.5–2 cm. *Bracts* linear, 2–3 mm. *Flowers* white slightly pink flushed on outside, fragrant, 15–18 mm diameter. *Calyx* tube 4–5 mm long, lobes triangular, 1–2 mm. *Corolla* tube *c.*1 cm, lobes 5–6, ovate to oblong-ovate, 8 × 4 mm. *Capsules* purple-black, ellipsoid or subglobose, 0.8–1.2 cm × 5–12 mm.
DISTRIBUTION. South and South-west China and Taiwan.
HABITAT. Valleys and woods; 900–2,200 m.
FLOWERING TIME. Summer to autumn.

Plate 14. *Jasminum urophyllum*. Painted by Lilian Snelling for *Curtis's Botanical Magazine*, t. 148 (1951).

Section UNIFOLIOLATA DC.

All the species in this section have opposite, usually simple leaves with articulate petioles. The flowers are white; there is usually one seed per fruit, except in *Jasminum simplicifolium* which has two. [The red-flowered *J. beesianum* has simple leaves, but is now placed in section *Jasminum.*] The species of this section are found mainly around the Pacific from Australia to Tropical Asia, including Thailand, Malaysia, Cambodia, Laos, Burma (Myanmar), India, China and Tropical Africa. The type species is *Jasminum sambac.*

18. JASMINUM ADENOPHYLLUM

Section Unifoliolata DC.

Jasminum adenophyllum can be recognised by having its flowers on very long pedicels and by its long calyx lobes. It is cultivated in Thailand as an ornamental and for its edible young shoots. It has recently been introduced to Europe from Thailand by Guy Sissons of the Plantsman Nursery, now located in SW France. He describes it as being very free flowering with long running and twining stems, grey-green leaves and big, heavily-scented, flowers with 8–9 thin petals.

It is probably not frost-hardy, and requires a minimum temperature of 10°C. Its hardiness rating is G2, USDA zones 10–11.

The specific name *adenophyllum* refers to the glands on the leaves, from the Greek adenos, meaning a gland.

Jasminum adenophyllum Wall. ex C. B. Clarke, *Fl. Brit. Ind. (J. D. Hooker)* 3: 597 (1882). Type: Habitat in Sylhet mountains, *Wallich* 2876 (K).

DESCRIPTION. Evergreen woody, glabrous or slightly hairy climber. *Branches* rather slender, terete, glabrous. *Leaves* opposite, simple, hairy on under surface, elliptic to oblong-elliptic, 6–15 × 2.5–7 cm, with 4–5 pairs of veins with tufts of hairs in the axils; base cuneate; apex shortly acuminate; petiole *c.*4–7 mm long, sometimes jointed. *Inflorescence* lax, terminal on side shoots or axillary, 1–5 flowered; pedicels 1–4 cm. *Bracts* linear, 2 mm. *Flowers* white, fragrant to 3–4 cm diameter. *Calyx* tube 2 mm long, lobes filiform, 5–15 mm. *Corolla* tube 9–20 mm, lobes 8–9, lanceolate to linear, 15–20 × 2–5 mm. *Berries* ellipsoid, *c.*10 mm.
DISTRIBUTION. India, Vietnam, and Thailand to Malaya.
HABITAT. Dry evergreen forests, bamboo thickets and scrub; 0–500 m.
FLOWERING TIME. Spring to late summer.

19. JASMINUM DECUSSATUM

Section Unifoliolata DC.

Jasminum decussatum is a hairy plant with ribbed leaves and white flowers, each with 7–8 corolla lobes, closely related to *J. scandens* (p. 113), and sometimes, e.g. by Green (2003) considered merely a hairy form of it. It is grown in Thailand as an ornamental, and, like *J. adenophyllum*, has recently been introduced to Europe from Thailand by Guy Sissons of the Plantsman Nursery in SW France.

It is a tropical species, requiring frost-free conditions: its hardiness rating is G2, USDA zones 10–11.

The specific name, *decussatum*, refers to the leaves or inflorescence branches being in pairs at right angles to one another.

Jasminum decussatum Wall. ex G. Don, *Gen. Syst.* 4: 62 (1837). Type: *Wallich* 2860 (K–W).

DESCRIPTION. Evergreen villous climber. *Branches* very hairy. *Leaves* opposite, simple, thick, ovate to lanceolate, 5–11 × 2–6 cm, with 3–4 pairs of veins; base rounded to truncate; apex acute to acuminate; petiole 6–16 mm long, jointed. *Inflorescence* terminal on side shoots, panicles, many flowered, pedicels 0–3 mm; flowers. *Bracts* linear, 1–10 mm. *Flowers* white, fragrant, 1.5–2 cm diameter. *Calyx* hairy, tube 1–1.5 mm long, lobes linear, often reflexed, 1–2 mm. *Corolla* tube 4–10 mm, lobes 7–8, lanceolate, acute to acuminate, 7–9 × 1–2 mm. *Berries* dark purple, ellipsoid, *c.*10 × 6 mm.

DISTRIBUTION. Burma, Malayan Peninsula, Sumatra and Thailand.

HABITAT. Evergreen forest, scrub and roadsides; to 700 m in Thailand.

FLOWERING TIME. Winter to spring.

20. JASMINUM DICHOTOMUM

Section Unifoliolata DC.

This plant was introduced into cultivation in the USA by Dr D. Fairchild and has become naturalised in Florida where it is known as Gold Coast Jasmine. It can be recognised by its large, glossy, broad, leathery leaves, dense inflorescence and flowers with long corolla tubes; it bears some resemblance to *Jasminum laurifolium* f. *nitidum,* but has much shorter calyx lobes. In Africa the berries are eaten and the fresh leaves used for the treatment of ulcers.

The name *dichotomum* refers to the dichotomous branching of the inflorescence. It requires greenhouse conditions. Hardiness: G2, USDA zones 10–11.

Jasminum dichotomum Vahl, *Enum. Pl.* 1: 26 (1804). Type: *Smeathman* s.n. (BM).

ILLUSTRATIONS. Turrill, *Flora of Tropical East Africa, Oleaceae,* 22, f. 7 (1952); Menninger, *Flowering Vines of the World*, fig. 197 (1970).

DESCRIPTION. Evergreen, woody glabrous or minutely hairy, climber or scrambler. *Branches* terete. *Leaves* opposite, sometimes in whorls of 3 at nodes, simple, leathery, dark green, glossy, elliptic or ovate-elliptic, 3–10 × 1.5–6.5 cm, with 2–4 pairs of veins, the lower very prominent; base acute to rounded; apex acute to shortly acuminate; petiole 7–20 mm long, articulate. *Inflorescence* dense, terminal or axillary, to 50 or more flowered, pedicels 1–3 mm. *Bracts* linear, 3–4 mm. *Flowers* white, tinged red or reddish purple on outside, sweetly fragrant especially at night, 1.5–3 cm diameter. *Calyx* tube 2–3 mm long, lobes linear to narrow ovate, 0.5–2 mm. *Corolla* tube 1.5–2 cm, lobes 5–9, oblanceolate, 7–16 × 2–4 mm. *Berries* black, ellipsoid, 12–15 × 3–6 mm.

DISTRIBUTION. Tropical Africa, from the Sudan to Zambia.

HABITAT. Forest and forest margins; 1,050–1,800 m.

FLOWERING TIME. Intermittently all year round.

Fig. 28. **_Jasminum dichotomum_**. Line drawing by Dorothy R. Thompson from Turrill, _Flora of East Tropical Africa_ 22. f.7 (1952).

21. JASMINUM ELONGATUM

Section Unifoliolata DC.

A widespread and very variable tropical jasmine from northern Australia through the Pacific to southeast Asia and China. It varies in its habit, degree of pubescence, length of calyx and corolla lobes, all of which are characters often used to separate species. However, Green (2000) considers that there are no clear discontinuities and the variation is due to its wide geographic range. In Australia it is grown as *Jasminum aemulum*, in the Himalayas as *J. amplexicaule* and in the Phillipines as *J. bifarium*. The form cultivated in Nepal is striking for the contrasting bright pink buds and white flowers (Herklots, 1976).

One of the earlier introductions into western gardens was as *Jasminum amplexicaule*, grown by Colvill's Nursery in King's Road, Chelsea, London by 1820. It is also close to *J. syringifolium* from Assam and Burma (Myanmar), which has a more open inflorescence and to *J. scandens* from India and Bangladesh, which has flowers with a shorter corolla tube.

This species tolerates very light frost and prefers a sunny position. It can be trained as a shrub or a climber and in Australia is said to be an excellent butterfly plant. The specific name refers to the elongated leaves. Its hardiness rating is G1–2, USDA zones 9–11.

2 cm

Fig. 29. **Jasminum elongatum**. Line drawing by Geoffrey Herklots. Nepal, 1961.

Jasminum elongatum (Bergius) Willd., *Sp. Pl.* ed. 41: 37 (1797). Type: China, Canton (SBT), *Ekeberg* s.n. (holotype SBT; photo. K).
Syn. *Nyctanthes elongata* Bergius, *Philos. Trans.* 61: 289, t. 11 (1772).
    J. aemulum R. Br., *Prodr.*: 521 (1810).
    J. amplexicaule Buch.-Ham. ex G. Don, *Gen. Syst.* 4: 60 (1837).
    J. bifarium Wall., ex G. Don, *Gen. Syst.* 4: 60 (1837).
    J. tonkinense Gagnep., *Fl. Gén. Indo-Chine* 3: 1053 (1933).
    J. undulatum (L.) Ker-Gawl., *Fl. Brit. India* 3: 592 (1882), non (L.) Willd.

ILLUSTRATIONS. *Allertonia* 3: (6) fig. 15 E & F (1984) as *J. aemulum*; *Botanical Register* 6: 436 (1820) as *J. undulatum*; Herklots, *Flowering Tropical Climbers*, p. 142, f. 205 (1976) as *J. amplexicaule*.
DESCRIPTION. Evergreen, straggly bush or strong climber. *Branches* round, hairy or almost glabrous. *Leaves* opposite, simple, bright green, rather thin, usually ovate to narrow lanceolate, 2–11 × 1.5–6 cm, with 2–4 pairs of prominent veins; base obtuse to rounded to subcordate; apex acute to

acuminate; petiole articulate, 3–10 mm long, often pubescent on the veins beneath. *Inflorescence* terminal on side shoots, sometimes hairy, dense, 3–17 flowered, pedicels 0.5–3 mm. *Bracts* 1–2 pairs of leafy bracts, 0.5–1 cm. *Flowers* white, fragrant, 2–3 cm diameter. *Calyx* finely hairy, tube 1.5–2 mm long, lobes filiform, 2–11 mm. *Corolla* tube 13–25 cm, lobes 7–9, narrow lanceolate to elliptic, acute, 6–12 × 2–4 mm. *Berries* black, ellipsoid, *c*.8 × 5 mm.

DISTRIBUTION. Northern Australia, Papua New Guinea, Indonesia, Phillipines, Malaysia, Thailand, Vietnam, China (Yunnan, Guizhou, Guandong), Burma, India (Assam), Bhutan, Sikkim.

HABITAT. Coastal woodlands, monsoon forests and open ground; from sea level to about 1,500 m.

FLOWERING TIME. Winter, spring and summer; April to December in Australia.

Related species:

21a. Jasminum perissanthum P. S. Green, *Kew Bull*. 50: 578 (1995). Type: Thailand, *Kerr* 4924 (holotype K, isotype L).

ILLUSTRATION. *Kew Bulletin* 50: 579 (1995).

This recently-named and beautiful jasmine was described from a specimen collected in 1921 by A. F. G. Kerr in northern Thailand. It was growing in evergreen forest at 1,600–1,700 m, at Nan, Doi Pu Ka, and flowering in late February. The leaves are simple, broadly lanceolate and pinnately veined, 4–5 cm long; the inflorescence consists of dense 10–20-flowered clusters of scented flowers, with the tube 25–30 mm long, the lobes 12–15 mm long. 'It is remarkable that this species, which must be very handsome in flower, is represented by only one collection amongst the very many specimens of *Jasminum* I have examined from Thailand' (Green, 1995). (Note: the specific name was derived from the greek perissos, meaning abundant. It was incorrectly spelled *perrisanthum* in the description, but, correctly, *perissanthum* in the caption.)

Fig. 30. *Jasminum perissanthum*. Line drawing by Judi Stone for *Kew Bulletin* 50: 579 (1995).

Plate 15. *Jasminum elongatum*. Hand-coloured engraving by Sydenham Edwards from *Botanical Register* 6, t. 436 (1820).

22. JASMINUM HARMANDIANUM

Section Unifoliolata DC.

Jasminum harmandianum is a woody climber which can be recognised by its very long and narrow corolla lobes. It is cultivated in Thailand where it is used in religious ceremonies in Buddhist temples. Normally the length of the calyx lobes is a useful diagnostic character but occasionally in this species the length is variable.

In frosty climates this species needs greenhouse protection; its hardiness rating is G2: USDA zones 10–11.

Jasminum harmandianum Gagnep., *Bull. Soc. Bot. France* 80: 74 (1933). Type: Delta du Mèkong, *Harmand* 633 (P).

DESCRIPTION. Evergreen woody, glabrous climber. *Branches* terete. *Leaves* opposite, simple, narrow oblong to elliptic, 5–12 × 2–4 cm long, with 5–6 pairs of veins; base acute to obtuse; apex shortly acuminate; petiole 4–6 mm long, jointed. *Inflorescence* lax, terminal on side shoots, panicles, 5–15 flowered, pedicels 2–10 mm. *Bracts* linear to lanceolate, 5–15 mm. *Flowers* white, often flushed pink on the outside, strongly fragrant, 2–3.5 cm diameter. *Calyx* tube 1.5–2 mm long, lobes subulate, 5–10 mm. *Corolla* tube 18–22 mm, lobes 7–8, narrow, long acuminate, 10–18 × 1–2 mm. *Berries* globose, *c*.5 mm.

DISTRIBUTION. Cambodia, Laos, Vietnam and Thailand.

HABITAT. Evergreen forest and margins, shrub vegetation; to 700 m.

FLOWERING TIME. Spring to summer.

23. JASMINUM KEDAHENSE

Section Unifoliolata DC.

Jasminum kedahense can be recognised by its large flowers in a dense head, good scent, short pedicels and very long corolla tubes. It was in cultivation at the Royal Botanic Gardens, Kew for many years before being illustrated in *Curtis's Botanical Magazine* in 1970. Although the original Kew records have been lost, it was possibly grown from a collection by H. N. Ridley in 1893. It is cultivated in several countries of the Far East, and was introduced from Singapore to Hong Kong by Dr Herklots in 1935. A well drained rich compost is best for this species and under glasshouse conditions it needs little pruning. It requires frost-free conditions and its hardiness is rated G2, USDA zones 10–11.

The specific name refers to the region of Kedah in north-west Malaysia, where this species was collected.

Jasminum kedahense (King & Gamble) Ridl., *J. Fed. Malay. States Mus.* 7: 46 (1916). Type: Malay peninsula, Kedah Peak, Dec. 1915, *Robinson & Kloss* 5981, annotated as type by Ridley (K). Syn. *J. maingayi* C. B. Clarke var. *kedahense* King & Gamble, *J. Asiat. Soc. Bengal, Pt. 2, Nat. Hist.* 74 (2): 258 (1906).

Plate 16. *Jasminum kedahense*. Painted by Margaret Stones for *Curtis's Botanical Magazine*, t. 547 (1969).

ILLUSTRATION. *Curtis's Botanical Magazine* ns. vol. 177: t. 547 (1970).

DESCRIPTION. Evergreen climber, roughly hairy when young. *Branches* stout, often hairy. *Leaves* opposite, simple, leathery, 5–17 × 2.5–7 cm long, broad ovate to ovate, often hairy below, with 4 main pairs of veins, the lower pair forming an indistinct submarginal vein; base obtuse to rounded to subcordate; apex acuminate; petiole 6–20 mm long, articulate. *Inflorescence* dense, axillary or terminal on side shoots, 6–15 or more flowered, pedicels 2–5 mm; flowers 3–4 cm across. *Bracts* linear, 3–6 mm. *Flowers* white, sometimes tinged red on outside, fragrant, 3–4 cm diameter. *Calyx* hairy, tube 3–4 mm long, lobes subulate, 5–7 mm. *Corolla* tube 2.5–3.5 cm, lobes 7–10, lanceolate, 15–23 × 4–6 mm. *Berries* black, ellipsoid, *c.*15 × 10 mm.

DISTRIBUTION. Thailand and Malay peninsula.

HABITAT. Evergreen forest and open scrub; 800–1,700 m.

FLOWERING TIME. Winter to spring.

24. JASMINUM LAURIFOLIUM forma NITIDUM

Section Unifoliolata DC.

Jasminum laurifolium usually forms a shrub, but also sends out long trailing shoots. It has shining evergreen leaves and flowers with numerous (usually 9–11) petals and very narrow sepals held at right angles to the corolla tube.

It flowered for the first time at Colvill's Nursery in Chelsea, London in 1820 where it was known as *Jasminum angustifolium* var. *laurifolium*. In 1899 it was listed by the nurseryman, W. Bull, as *Jasminum nitidum*, having been recently introduced as a new species reputedly from the Admiralty Islands; however this is unlikely as it has never since been found in wild there despite several attempts to find it (Green, 1984).

Jasminum laurifolium var. *laurifolium* comes from the hills of eastern Assam and eastern Bangladesh (with one record from Thailand) and is probably not in cultivation. The species is variable in leaf width, but f. *nitidum* is distinguished by its broader leaves and very narrow, finely hairy calyx lobes. A second variety, var. *brachylobum* Kurz, differs in its very small calyx lobes, only 2–3 mm long and sometimes broader leaves. It is more widespread in the wild, growing in bamboo forest in SW Thailand, Burma and southern China from Yunnan to Hainan.

At first, *Jasminum laurifolium* f. *nitidum* grows as a bush and it may be pruned to keep it more bush-like, but later it sends out long twining stems to three metres or more. It is a useful pot plant; it flowers while still quite small and will cope in a house as it survives low light intensity. In tropical regions, and in warm climates such as southern California, it may be used for ground cover or as an informal hedge but in these conditions it prefers full sun.

The large numerous star-like flowers with very narrow corolla lobes make it a popular garden plant in many tropical climates and it is known by numerous common names including angel wing, angel hair, windmill, and star jasmine (especially in the USA). There is also a variegated form with a yellow splashed centre to the leaf in cultivation in the USA. In frosty climates it requires protection; hardiness G2: USDA zones 9–11.

Jasminum laurifolium Roxb., *Fl. Ind.* 1: 91 (1820), f. **nitidum** (Skan) P. S. Green, *Kew Bull.* 39: 655 (1984). Type: Cultivated at Chelsea, by W. Bull and Sons (K).
Syn. *J. nitidum* Skan, Bull. Misc. Inform., Kew 1898: 225 (1898).

Fig. 31. *Jasminum laurifolium* forma *nitidum*. Line drawing by Geoffrey Herklots.

2 cm

Fig. 32. *Jasminum laurifolium* forma **nitidum**.

ILLUSTRATIONS. Graf, *Tropica* edn. 3, 697 (1986). Herklots, *Flowering Tropical Climbers*, fig. 207, p. 138 (1976) as *J. nitidum* Skan; *Botanical Register* 7: 521 (1821) as *J. angustifolium* (L.) Willd. var. *laurifolium* (Roxb.) Ker-Gawl.; Menninger, *Flowering Vines of the World*, fig. 201 (1970); Ellison, *Cultivated Plants of the World*, p. 329 (1995); Rix, *Subtropical and Dry Climate Plants*, p. 114 (2006).

DESCRIPTION. Evergreen weak climber, or branching shrub to *c.* 1.5 m. *Leaves* opposite, simple, glossy, bright dark green, leathery, lanceolate or ovate, to 10 × 2–3 cm long, with the basal pair of veins forming a sub-marginal vein; base sub-cordate to rounded; apex shortly acuminate; petiole articulate, 5 mm long. *Inflorescence* lax axillary and terminal, *c.* 5 flowered, pedicels 10–18 mm. *Bracts* subulate, *c.*4 mm. *Flowers* white, often tinged purplish red on the outside, fragrant, 2.5–4 cm diameter. *Calyx* hairy, tube 2–3 mm long, lobes filiform radiating at right angles to the corolla tube sometimes flushed green, to 10 mm. *Corolla* tube 2 cm, lobes 6–12, narrowly lanceolate, 20 × 3 mm. *Berries* not seen.

DISTRIBUTION. This form is apparently only known in cultivation, and is of uncertain origin.

FLOWERING TIME. Late spring to autumn.

25. JASMINUM MAINGAYI

Section Unifoliolata DC.

This species is similar to *Jasminum kedahense* but has smaller flowers and larger but narrower leaves.

This species has been grown at Kew since the beginning of the twentieth century. It is named after Dr A. C. Maingay who first found the plant, but was killed quelling riots in Rangoon in 1869. In frosty climates it requires protection; its hardiness rating is G2, USDA zones 9–11.

Plate 17. *Jasminum maingayi*. Hand-coloured lithograph by Matilda Smith from *Curtis's Botanical Magazine*, t. 7823 (1902).

Jasminum maingayi C. B. Clarke, *Fl. Brit. Ind.* 3: 594 (1882). Type: Malaya, Penang: *Maingay 1000* (K).

ILLUSTRATION. *Curtis's Botanical Magazine* 58: t. 7823 (1902).

DESCRIPTION. Evergreen glabrous or hairy climber. *Branches* hairy. *Leaves* opposite, simple, dark green, glossy, somewhat leathery, narrow ovate to oblong ovate, with 3–5 pairs of veins of main veins, 9–17 × 2.5–8 cm; base obtuse attenuate into petiole; apex acuminate; petiole 3–15 mm long, articulate. *Inflorescence* dense, terminal on side shoots, 7 to many-flowered, pedicels erect, 2–5 mm. *Bracts* linear, 3–5 mm. *Flowers* white, sometimes with faint reddish flush on outside, fragrant, to 1.5–2.5 cm diameter. *Calyx* tube 2–3 mm long, lobes lanceolate, 5–10 mm. *Corolla* tube 15–25 mm, lobes 8–10, narrow oblong, 8–15 × 3– 6 cm. *Berries* ellipsoid, *c.*7 × 4 mm.

DISTRIBUTION. Penang, Malaysia and Thailand.

HABITAT. In evergreen forest, and on rocky slopes; usually above 200 m.

FLOWERING TIME. Autumn to spring.

26. JASMINUM MOLLE

Section Unifoliolata DC.

This is one of the twelve species of Jasmine native to northern Australia, and is found from the Kimberleys to the Gulf of Carpenteria. It is cultivated there and is useful in tropical areas with a marked dry season; it is also currently sold by Guy Sissons of the Plantsman Nursery in SW France and probably by other nurserymen in Europe. It is an attractive plant grown for its bushy habit and sweetly scented heads of many flowers. It may be trained as a shrub or allowed to climb, but it is slightly slower growing than many other related species. An open sunny position in well-drained soil is best for this plant and it will survive fairly dry conditions. It is close to *Jasminum simplicifolium* subsp. *australiense,* differing in the lack of an obvious marginal vein on its softly hairy simple leaves. It may also be confused with the very hairy *J. decussatum* but differs in the generally smaller leaves and more spreading inflorescence.

Jasminum molle R. Br., *Prodr. Fl. Nov. Holland.* 521 (1810). Type: Northern Territory, South Bay Point (Melville Bay) 1803, *R. Brown 2843* (BM).

ILLUSTRATION. *Allertonia* 3: (6) fig. 15 E & F (1984).

DESCRIPTION. Evergreen straggling shrub or twining climber hairy, especially when young. *Branches* terete, usually hairy but occasionally glabrous. *Leaves* opposite, simple, broad ovate to lanceolate, 2–10 × 1–6 cm long, with 2–3 prominent pairs of veins; base cuneate attenuate to petiole; apex rounded; petiole jointed at base, 5–15 mm long. *Inflorescence* terminal on side shoots, 2–5 cm, 9–many flowered, pedicels 1–20 mm. *Bracts* linear, 1–2 mm. *Flowers* white, fragrant, to 1.5–2 cm diameter. *Calyx* tube 1 mm long, lobes short, 0.5–1 mm. *Corolla* tube 8–16 mm, lobes 5–6, lanceolate, acute, 6–10 × 2.5–3.5 mm. *Berries* purplish-black, globose, *c.*12 mm.

DISTRIBUTION. Northern Australia (endemic).

HABITAT. Mixed open forest and woodland in sandy soils; *c.*0–300 m.

FLOWERING TIME. Late summer.

Two other Australian species in section Unifoliata may be cultivated in Australia:

26a. Jasminum calcareum F. Muell., *Fragm.* 1: 212 (1859). Type: Western Australia *Oldfield* 1859 (MEL).

This species is endemic to western Australia and the Northern Territories where it grows on limestone in semi-arid regions, and is particularly common in the ranges near Alice Springs. It has the common name poison jasmine, as the leathery leaves are said to be poisonous to stock. The flowers have a pink tube and around eight narrow lobes. It is described and illustrated in Elliot & Jones, *Encyclopaedia of Australian Plants Suitable for Cultivation*, (vol. 5, p. 475), where it is reported to be tolerant of frost and useful for dry conditions.

26b. Jasminum kajewskii C. T. White, *Contr. Arnold Arbor.* 4: 90 (1933). Type: *Kajewski* 1346 (BRI).

This vigorous climber is endemic to the rainforests of Queensland where it grows into the tops of trees. It has large, fragrant flowers around 4 cm across with 6–7 narrow lobes; its glossy, few veined leaves are broadly ovate to ovate-lanceolate, 2–6 cm across. It is described in Elliot & Jones, *Encyclopaedia of Australian Plants Suitable for Cultivation*, (vol. 5, p. 476).

27. JASMINUM MULTIFLORUM

Section Unifoliolata DC.

This is one of the most widely cultivated species, grown in the tropics throughout the world for its pure white star-like flowers and sweet scent, which is best in the morning. It is commonly known as the star jasmine. It can be recognised by its softly pubescent shoots, very long calyx lobes and leaves which are broad, short-stalked and truncate at the base; the whole inflorescence is congested and leafy.

As well as being grown as an ornamental, the dried corollas are used for making tea and the fresh flowers for leis in Hawai'i. It has been cultivated since ancient times in India and neighbouring countries but has been known in western gardens only since the time of Phillip Miller in 1759. There appear to be several clones in cultivation, not all of which are equally fragrant (Green, 1965). Some of these in the past have been given distinct specific epithets and plants in cultivation as *Jasminum undulatum* usually belong to this species.

This species tolerates some shade and heavy pruning and in suitable climates can be used as ground cover or allowed to fall over a bank where it will flower all year round. Elsewhere it may be grown in a container. It requires protection in frosty climates; hardiness G2, USDA zones 9–11.

Jasminum multiflorum (Burm. f.) Andrews, *Bot. Repos.* 8: t. 496 (1807). Type: described from India.
Syn. *Nyctanthes multiflorum* Burm. f., *Fl. Ind.* 5, t. 3, fig.1 (1768).
    J. hirsutum (L.) Willd., *Sp. Pl.* 1: 36 (1797).
    J. pubescens (Retz.) Willd., *Sp. Pl.* 1: 37 (1797).
    J. gracillimum Hook. f., *Gard. Chron.* 1: 9 (1881).

ILLUSTRATIONS. Herklots, *Flowering Tropical Climbers*, p. 142, f. 206 (1976); *Curtis's Botanical Magazine* 45: t. 1991 (1818) as *J. hirsutum* and 37: t. 6559 (1881) as *J. gracillimum*; Menninger, E, *Flowering Vines of the World*, fig. 196 (1970).

2 cm

Fig. 33. *Jasminum multiflorum*. Line drawing by Geoffrey Herklots, Hong Kong, 1946.

DESCRIPTION. Evergreen densely hairy, weak climber. *Branches* round. *Leaves* opposite, simple, hairy below and above, lanceolate-oblong to broadly ovate, 3–8 × 1.5–5 cm, with about 4 pairs of veins, base truncate to sub-cordate; apex acute to acuminate; petiole 5–15 mm long, articulate. *Inflorescence* in dense terminal cymes on side shoots, panicles, 5–15 or more flowered, pedicels 0–2 mm. *Flowers* white, fragrant, to 2.5–4 cm diameter. *Bracts* basal ovate, 1.5–2 cm, upper linear, 3–5 mm. *Calyx* densely hairy, tube *c*.1 mm; lobes filiform, to 4–10 mm. *Corolla* tube often tinged greenish–yellow, 12–17 × 4–5 mm, lobes 7–9, narrowly elliptic, overlapping, 12–16 × 3–5 mm. *Berries* black, ellipsoid, *c*.1 cm long (not always produced in cultivation).

DISTRIBUTION. Tropical Asia. India to China but widely naturalised in tropical Asia.

HABITAT. Open thickets in forest; to *c*.600 m.

FLOWERING TIME. Mainly in autumn, extending to winter and spring.

28. JASMINUM MULTIPARTITUM

Section Unifoliolata DC.

This species is known as starry jasmine and is grown in gardens in southern Africa where it tolerates mild frost and semi-shade. The leaves are quite small but the flowers are large, with a very long corolla tube and calyx lobes, and have a strong sweet fragrance, especially at night.

There appears to be two forms in cultivation in South Africa. One has larger flowers to 4 cm across, which are borne singly and scattered over the whole plant. The leaves are slightly larger but the habit is more bushy than climbing. This can easily be treated as a shrub if any long branches are removed. The plant shown on plate 19 belongs to this form.

Plate 18. *Jasminum multiflorum* **'Gracillimum'**. Hand-coloured lithograph by Anne Barnard from *Curtis's Botanical Magazine*, t. 6559 (1881).

The second form is more of a climber, although not fast growing, and is more floriferous, although the individual flowers and leaves are smaller. This form cannot be successfully trained as a shrub. Both can be grown in large pots in frosty areas, and brought under cover and kept rather dry in winter. The species will tolerate light frost: hardiness G1: USDA zones 8–11.

Jasminum multipartitum is also used as a love charm, an emetic, in pot-pourri, and in herbal tea. The foliage is browsed by animals and the fruits are eaten by birds and even by people in famine times.

Jasminum multipartitum Hochst., *Flora* 27: 825 (1844). Type: Natal, Durban bay, woods near bay, *Krauss* 458 (K).
Syn. *J. glaucum* (L. f.) Aiton var. *parviflorum* E. Mey.

ILLUSTRATIONS. *Botanical Register* 23: 2013 1837 as *Jasminum glaucum* is probably this species (Green, 1963); *Flowering Plants of South Africa* 32: 1272 (1958); Kupicha, in *Flora Zambesiaca* 7 (1): tab. 71. (1983).

DESCRIPTION. Evergreen bushy scrambling plant but not a strong climber. *Branches* with very short inconspicuous hairs. *Leaves* opposite, simple, bright dark green, glossy, oblong to ovate-oblong or ovate-lanceolate, 1.5–5 × 0.4–5 cm long, with 3–4 pairs of obscure veins, the lower pair more prominent, base rounded; apex acute to rounded; petiole 2–4 mm long, articulate. *Inflorescence* solitary, or occasionally to 3, on short side shoots; pedicels 1–5 mm. *Bracts* linear, to 5 mm or more. *Flowers* white, flushed pinkish-red outside, buds maroon-red; fragrant, to 4 cm diameter or more. *Calyx* glabrous or shortly hairy, tube 2–3 mm long, lobes linear, 3–7 mm. *Corolla* tube 3–5 cm, lobes 7–11, lanceolate to elliptic, to 15–30 × 3–8 mm. *Berries* black, glossy, ellipsoid, *c.*15 × 10 mm.

DISTRIBUTION. South Africa (Eastern Cape to Transvaal), Swaziland north to Zimbabwe and Mozambique.

HABITAT. Woodland, thickets, scrubland and on rocky slopes.

FLOWERING TIME. Spring to early summer.

A number of other South African species may occasionally be found in gardens there, particularly those devoted to their native flora, but they are rarely, if ever, grown elsewhere. For example:

28a. Jasminum glaucum (L. f.) Aiton, *Hortus Kew.* 1: 9 (1789), is found in the south west Cape, from the Gifberg to the Riviersonderend mountains, growing on riverbanks and rocky slopes. It is a loose shrubby plant with leathery, glaucous, lanceolate to elliptic leaves over 3 times as long as wide and large flowers with 5–7 rather broad corolla lobes, produced in late spring. It is very drought and heat tolerant.

28b. Jasminum streptopus E. Mey. var. **transvaalensis** (S. Moore) I. Verd., *Bothalia* 6: 572 (1956) from eastern South Africa, in the Natal coast and Midlands, is hairy and the leaves have tufts of hairs in the axils of the veins. The flowers are held on long slender pedicels.

28c. Jasminum stenolobum Rolfe in Oates, *Matabeleland* ed. 2: 403 (1899) is similar to *J. streptopus* but is distinguished by short thick pedicels and the leaves are without tufts of hairs in the axils of the veins. It is found in the coastal region and Midlands of Natal. This species has been confused with *J. multipartitum* in collections but is conspicuously pubescent.

Plate 19. *Jasminum multipartitum*. Painted by Christabel King for *Curtis's Botanical Magazine*.

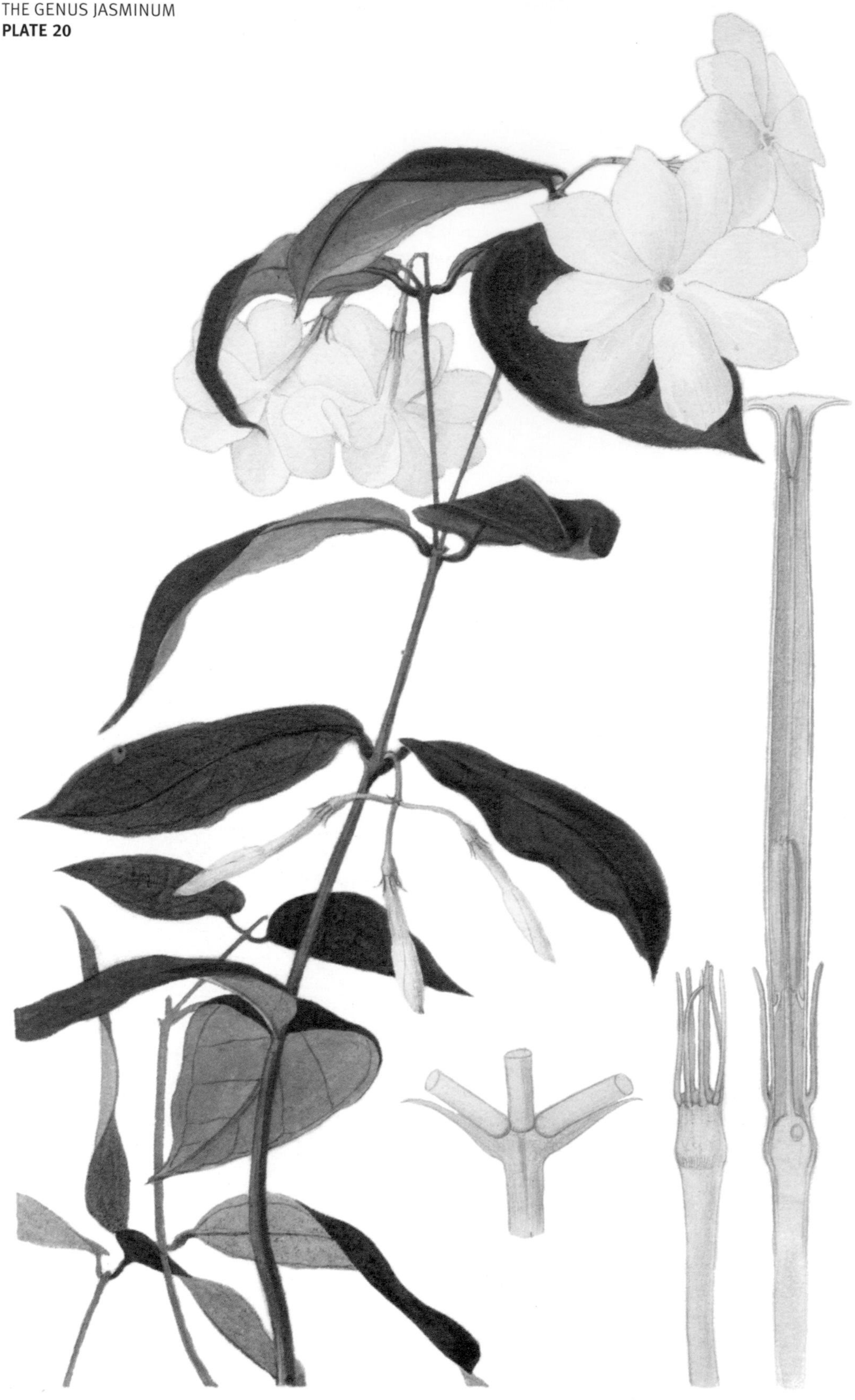

29. JASMINUM NOBILE subsp. REX

Section Unifoliolata DC.

This subspecies was described as *Jasminum rex* from Thailand in 1921 and, in 1995, combined as a subspecies with *J. nobile*. It is the largest-flowered white jasmine in cultivation and has unscented flowers to 7.5 cm across with broad overlapping petals. The buds may be pink-tinged.

The first specimens were sent to Kew in 1882 by H. J. Murton, head gardener to the King of Siam who collected it on the mountain Khao Sai Dao (Kao Soi Das) near the Thailand-Cambodia border on the Malay Peninsula, but the first living plant was sent in a Wardian case in 1921 by Mr Sanitwongse who grew it on his pergola in Bangkok. In a letter to the *Gardeners' Chronicle* in December 1938, A. F. G. Kerr recorded that it grows in the lowland forest east and northeast of Chantabun (probably Chanthaburi), near the type locality. It flourished in the stove house at Kew where it flowered a year later. It remains a popular plant, widespread in gardens throughout the tropics and also in cultivation under glass in many temperate regions. The flowers are the largest of the genus and formed more or less throughout the year. It has no special cultural requirements, but does need tying up at an early stage because it is not a strong climber. It requires protection from frost; hardiness G2: USDA zones 10–11.

The subspecies *nobile* from near Moulmein in Burma is probably not in cultivation but has slightly smaller flowers and broader leaflets.

Jasminum nobile C. B. Clarke subsp. **rex** (Dunn) P. S. Green, *Kew Bull.* 50 (3): 577 (1995). Type: Bangkok, cultivated 1921, *Sanitwongse* s.n. (K).
Syn. *J. rex* Dunn, *Bull. Misc. Inform., Kew* 1921: 219 (1921).

ILLUSTRATIONS. *Curtis's Botanical Magazine* 148: 8934 (1922) (as *J. rex*); *Gardener's Chronicle* 26: 11 (1938); Herklots, *Flowering Tropical Climbers*, p. 143, f. 208 (1976); Menninger, *Flowering Vines of the World*, fig. 203 (1970); Ellison, *Cultivated Plants of the World*, p. 330 (1995).

DESCRIPTION. Woody, evergreen, glabrous weak climber. *Leaves* opposite, simple, somewhat leathery, glossy, dark green,

2 cm

Fig. 34. *Jasminum nobile* subsp. *rex*. Line drawing by Geoffrey Herklots.

Plate 20. *Jasminum nobile* subsp. *rex*. Painted by A. Kellet for *Curtis's Botanical Magazine*, t. 8934 (1922)

oblong-elliptic or ovate to broad-lanceolate, 5–14 × 2–5 cm long, with 2 main marginal veins, rest obscure; base acute to obtuse; apex acuminate; petiole 8–10 mm long, articulate. *Inflorescence* terminal on side shoots, 1–3-flowered or more; pedicels 20–40 mm. *Bracts* linear, 2–5 mm long. *Flowers* white, sometimes flushed purplish-red outside, not fragrant, to 7.5 cm diameter. *Calyx* tube 3–4 mm long, lobes subulate, 5–10 mm. *Corolla* tube 25–35 mm, lobes 8–9, oblong-elliptic, broad, overlapping, 23–30 × 10–18 mm. *Berries* black, ellipsoid, *c.*12 × 8 mm.

DISTRIBUTION. Thailand, mainly in the southeast, and Cambodia.

HABITAT. Evergreen forests in lowlands.

FLOWERING TIME. Intermittently all year round.

Jasminum 'Ann Clements' (*J. nobile* subsp. *rex* ×*multiflorum*), was first registered in 1959. This is the result of repeated crosses by Paul Swedroe of Fort Lauderdale, Florida, to try to introduce scent into a large-flowered jasmine. It is a vigorous plant with large, prolific, scented flowers and dark foliage. It has however never become widely successful in the nursery trade, perhaps because it is less easy to grow.

ILLUSTRATION. Menninger, *Flowering Vines of the World*, fig. 194 (1970).

30. JASMINUM SAMBAC

Section Unifoliolata DC.

This species has been cultivated for centuries in tropical Asia, especially in China and India. Its first introduction to Europe is said by Clusius (1665) to have been to Florence, having been received from Cairo in 1660.

It is widely used for religious ceremonies, the perfume industry, for tea, as ornamental leis and garlands and in traditional medicine as well as an ornamental for its highly scented flowers.

A number of cultivars have been named in different parts of the world and several have become confused in cultivation. The most common are the very double-flowered 'Grand Duke of Tuscany' and the semi-double flowered 'Maid of Orléans'. Not all cultivars are climbing and most make attractive bushy plants suitable for growing as house plants. This species enjoys a rich loamy, slightly acid, soil and will tolerate dry conditions as well as a temporary slight frost. Hardiness rating G2, USDA zones 10–11. The species name is derived from the Persian vernacular name, *zambac* (see p. 9).

Jasminum sambac (L.) Aiton, *Hort. Kew.* 1: 8 (1789). Type: Habitat in India. Herb. Linn. Syn. *Nyctanthes sambac* L.

ILLUSTRATIONS. *Botanical Register* 1: t. 1 (1815); Graf, *Exotica* edn. 3, 1197 (1963); Everett, *New York Botanical Garden Illustrated Encyclopedia of Horticulture* 6: 1851 (1981); Herklots, *Flowering Tropical Climbers,* p. 144, f. 209 (1976); Ellison, *Cultivated Plants of the World,* p. 330 (1995).

DESCRIPTION. Vigorous evergreen hairy suberect shrub or weak climber to about 2 m. *Branches* round or angled. *Leaves* opposite, sometimes in 3's, simple, glossy, dark green, glabrous except for tufts of

Plate 21. *Jasminum sambac*. Hand-coloured engraving by Sydenham Edwards for *Botanical Register* 1, t. 1 (1815).

hairs in vein axils, elliptic to broadly ovate, to 9 × 6 cm, with 4–6 pairs of veins; base cuneate or rounded; apex acute or obtuse; petiole 2–6 mm long. *Inflorescence* terminal on side shoots, 1–10-flowered, pedicels to 6 mm or more. *Bracts* linear, to 4–8 mm. *Flowers* white, often fading to pinkish as they age, with very heavy fragrance, especially in the morning, 3–5 cm diameter. *Calyx* tube 2–3 mm long, lobes linear, 5–9 mm. *Corolla* tube 1–1.2 cm, lobes very variable, oblong to rounded, 8–12 × 5–9 mm. *Berries* purple-black, globose, to 10 mm (usually only produced on single flowered forms).

DISTRIBUTION. Tropical Asia, probably originally from India, but naturalised in other regions such as Arabia.

HABITAT. Sea level to 600 m.

FLOWERING TIME. Late spring to late summer.

The following cultivars are grown in Europe:

'Asian Temple'
A bushy, less vigorous double-flowered form.

'Bangkok Peony'
This cultivar has rather loose double flowers. It was found in a Bangkok market by Guy Sissons.

'Belle of India'
This good scented double-flowered cultivar has narrower slightly twisted petals and is grown in India. Although it shows a tendency to climb, it may be grown as a bush by pruning. It is slightly faster growing than 'Grand Duke of Tuscany'.

'Grand Duke of Tuscany'
A bushy, slow growing shrub, with the leaves sometimes arranged in threes at each node, and with very double, almost hemispherical flowers with round corolla-lobes. This is also sometimes grown incorrectly as 'Grand Duke' or 'Duke of Tuscany'.

ILLUSTRATIONS. *Curtis's Botanical Magazine* 43: 1785 (1816); Menninger, *Flowering Vines of the World*, fig. 199 (1970); Ellison, *Cultivated Plants of the World*, p. 330 (1995).

'Little Bo'
Introduced from Vietnam by Guy Sissons. It has a dwarf bushy habit, leaves with an undulate margin and lotus-shaped flowers produced in 2–3 flushes each year.

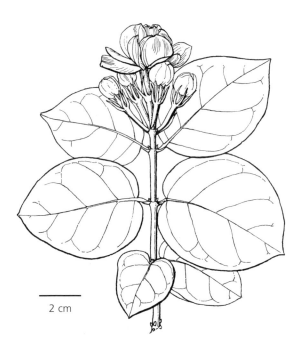

2 cm

Fig. 35. *Jasminum sambac* **'Grand Duke of Tuscany'**.
Line drawing by Geoffrey Herklots.

'**Maid of Orléans**' (named for Joan of Arc) AGM 2002.
This cultivar has semi-double extremely fragrant, long-lasting flowers.

'**Thai Beauty**'
This plant has a bushy habit and slightly undulate leaves. It has double hose-in-hose flowers with 3 or 4 tiers of rather pointed corolla lobes.

31. JASMINUM SCANDENS

Section Unifoliolata DC.

Jasminum scandens can be recognised by its small flowers with very long corolla tubes in relation to the flower size. It is grown in the Far East as an ornamental and also reported as cultivated in Britain in early nineteenth century. Guy Sissons records that in the wild in Thailand it can be seen growing by the side of the road in hedges, where it climbs and then cascades with a weeping habit. It requires protection from frost. Hardiness rating G2, USDA zones 10–11.

Jasminum scandens (Retz.) Vahl, *Symb. Bot.* 3: 2 (1794).
Syn. *Nyctanthes scandens* Retz., *Observ. Bot.* 5: 9 (1789).

DESCRIPTION. Shrub or woody evergreen somewhat hairy climber. *Leaves* opposite, simple, lanceolate, 3–12 × 1.5–6 cm, with 4 (–5) pairs of veins, hairy on undersurface; base rounded; apex acute to acuminate; petiole 5–20 mm long. *Inflorescence* dense, terminal on side shoots, 5–15 flowered, pedicels 1–3 mm. *Bracts* leaf-like, 10–30 × 6–10 mm. *Flowers* white, fragrant, 1–1.5 cm diameter. *Calyx* tube 1–2 mm long, lobes subulate to triangular, *c*.1 mm. *Corolla* tube 10–12 mm, lobes 5–7, oblong-elliptic, 5–8 × 1.5–2 mm. *Berries* black, ellipsoid *c*.10–11 × 7–8 mm.
DISTRIBUTION. India, Bangladesh, Burma and Thailand.
HABITAT. Evergreen forests.
FLOWERING TIME. Winter to spring.

32. JASMINUM SIMPLICIFOLIUM

Section Unifoliolata DC.

This is a variable species found in Australia and the Pacific Islands of which three subspecies are grown in gardens. *Jasminum simplicifolium* itself is probably not widely cultivated.

Jasminum simplicifolium G. Forst., *Fl. Ins. Austr.*: 3 (1786). Type: 'Amicorum Insulae' (Tonga) *Forster* (K).

1. Leaves linear to narrow elliptic . . . . subsp. *suavissimum*
1a. Leaves broadly ovate to lanceolate **2**
2. Corolla tube under 12 mm . . . . subsp. *australiense*
2a. Corolla tube to 20 mm, rarely under 12 mm . . . . subsp. *leratii*

Fig. 36. *Jasminum simplicifolium* subsp. *suavissimum*.

32a. JASMINUM SIMPLICIFOLIUM subsp. SUAVISSIMUM

This subspecies can be recognised by its very narrow linear or linear-elliptic leaves and sweetly scented flowers around 3 cm across, with 5–8 lanceolate lobes.

It was introduced to cultivation in the Arnold Arboretum, USA by Peter Green, who brought five seeds from Fiji in 1963. It is not widely cultivated except in Australia, but is tolerant of drought and very light frost. It needs a well-drained soil and full sun. Hardiness rating G1, USDA zones 9–11.

Jasminum simplicifolium G. Forst. subsp. **suavissimum** (Lindl.) P. S. Green, *Allertonia* 3 (6): 424 (1984). Type: Queensland, 18 Dec. 1846. Mitchell (CGE).
Syn. *J. suavissimum* Lindl. in T. Mitch., *J. Exped.Trop. Australia* 355 (1848).

ILLUSTRATIONS. *Allertonia* 3 (6): 420, f. 9A & B (1984); Phillips and Rix, *Conservatory and indoor plants* 2: 103 (1997).

DESCRIPTION. Slender shrub, trailing plant or weak, more or less glabrous climber. *Branches* slender. *Leaves* opposite, simple, bright green, glossy, linear or narrow elliptic, 1–7 × 0.2 or occasional to 1 cm; petiole 1–2 mm long; with an obscure submarginal and other obscure veins, base rounded into petiole; apex acute. *Inflorescence* terminal, lax, 1–9 flowered, pedicels 5–40 mm. *Bracts* linear, 1–2 mm. *Flowers* white, fragrant, 2–3 cm diameter. *Calyx* tube 1–2 mm long, lobes linear, 1–5 mm. *Corolla* tube 6–15 mm, lobes 5–8, lanceolate, acute, 6–11 × 1.5–2.5 mm. *Berries* black, glossy, ellipsoid, *c*.8 mm long.

Plate 22. *Jasminum simplicifolium* subsp. *australiense*. Hand-coloured engraving by Sydenham Edwards for *Curtis's Botanical Magazine*, t. 980 (1807).

Fig. 37. *Jasminum simplicifolium* subsp. *australiense* on Lord Howe Island. Photograph by Bill Baker.

DISTRIBUTION. Australia: eastern New South Wales and eastern Queensland.
HABITAT. Open forest, woodland, grassland and rocky slopes.
FLOWERING TIME. Spring to summer.

32b. JASMINUM SIMPLICIFOLIUM subsp. AUSTRALIENSE

This subspecies, which has sometimes been known as the wax jasmine, was introduced from Australia to Europe at the end of eighteenth century. It was described by Jacquin in his work illustrating the plants of the Palace of Schönbrunn (near Vienna) in 1798, as *Jasminum volubile*. It can be recognised by its open inflorescence with numerous starry, narrow-petalled flowers.

In England, it was grown by the Duke of Portland at Bulstrode, from where material was taken for the *Botanical Magazine* plate in 1807. At this time, except for *Jasminum sambac*, it was the only simple leaved jasmine known in cultivation. The scent of the flowers is less heavy than some species, but it is still grown in Europe, and in North America where a form listed as 'Maculata' has a central yellow variegation to the green leaves.

This subspecies is close to subsp. *leratii*, but can be distinguished by the shorter calyx lobes and corolla tube. It requires full sun, plenty of water and good drainage to flourish and will tolerate very slight frost. Slightly cooler winter temperatures will initiate the formation of the flower buds. Hardiness rating G1, USDA zones 8–11.

Jasminum simplicifolium G. Forst. subsp. **australiense** P. S. Green, *Allertonia* 3 (6): 419 (1984). Type: Queensland, North Gorge, 28 Oct. 1963, *N. H. Speck* 1924 (K).
Syn. *J. volubile* Jacq., *Pl. Rar. Hort. Schoenb.* 3: 39 (1798).
      J. gracile Andrews, *Bot. Repos.* t. 127(1800).

ILLUSTRATIONS. *Curtis's Botanical Magazine* 25: t. 980 (1806) as *J. simplicifolium*; *Botanical Register* 8: 606 (1822); Phillips and Rix, *Conservatory and Indoor Plants* 2: 103 (1997) as *J. volubile*; Menninger, *Flowering Vines of the World*, fig. 200 (1970).

DESCRIPTION. Shrubby glabrous climber. *Branches* terete. *Leaves* opposite, simple, leathery, bright but dark, glossy green, broad-ovate to lanceolate, 3–9 × 1–4 cm; petiole articulate, 5–10 mm long; lower pair of veins forming an obscure submarginal vein; base acute to rounded; apex acute to obtuse. *Inflorescence* terminal on side shoots, 10 cm, 10 to many flowered, pedicels 2–10 mm. *Bracts* 1–2 mm. *Flowers* white, sometimes green-tinged on the outside, fragrant, 1.5–2 cm diameter. *Calyx* tube 2 mm long, lobes triangular, to 1–3 mm. *Corolla* tube 8–12 mm, lobes 5–8, lanceolate, acute, 6–10 × 2–3 mm. *Berries* black, glossy, ellipsoid, *c*.12 mm long.

DISTRIBUTION. Western Australia, New Caledonia, Norfolk Islands.

HABITAT. Open forest and low altitude rainforest; 0–300 m.

FLOWERING TIME. Spring to summer.

32c. JASMINUM SIMPLICIFOLIUM subsp. LERATII

The history of this subspecies in gardens is uncertain. It has been in cultivation in California since the early part of the twentieth century, under a number of incorrect names, in Australia and almost certainly in France as well. It has been speculated that John Gould Veitch (who visited New Caledonia in 1865) might have collected it, but no records are available. Alternatively, it may have come via southern France as New Caledonia was a French colony at the time. In cultivation it is said to need good light and cool winter temperatures to ensure profuse flowering in spring. The leaf shape is variable and some of the younger leaves may be very narrow, approaching those of subsp. *suavissimum*.

The subspecific name commemorates the plant collector Le Rat.

Jasminum simplicifolium G. Forst. subsp. **leratii** (Schltr.) P. S. Green, *Flore de la Nouvelle Calédonie* 22: 54 (1998). Type: Magenta, Jan. 1903, *Le Rat 172* (B).
Syn. *J. leratii* Schltr, *Bot. Jahrb. Syst.* 40 *Beibl.* 92: 32 (1908).
   J. absimile L. H. Bailey, *Gentes Herb.* 4: 346 (1940).

DESCRIPTION. Evergreen minutely hairy or glabrous, scrambler or climber. *Leaves* glossy green, opposite, simple, very variable, ovate to lanceolate, sometimes broader or narrower, with 2–4 pairs of obscure veins with lower pair more prominent and marginal, 1–8 × 0.4–5 cm; base obtuse to cuneate to rounded or subcordate; apex obtuse or acute; petiole, articulate, 2–20 mm long. *Inflorescence* terminal or axillary on side shoots, lax, paniculate, 3–17-flowered, pedicels 2–20 mm. *Bracts* linear to filiform 1–5 mm. *Flowers* white, with variable fragrance, to 1.5–2.5 cm diameter. *Calyx* tube 1.5–3 mm long, lobes triangular-lanceolate, *c.* 1–3.5 mm. *Corolla* tube often greenish-white, 10–16 mm, lobes 5–7, lanceolate or narrower, acute, 5–12 × 2–3 mm. *Berries* black, ellipsoid-ovoid, 6–8 × 5–6 mm.

DISTRIBUTION. New Caledonia and Loyalty Islands.

HABITAT. Forest, dry scrub and savannah; to *c*.500 m.

FLOWERING TIME. Spring to summer.

33. JASMINUM SYRINGIFOLIUM

Section Unifoliolata DC.

Jasminum syringifolium is grown in the Far East, and old British gardening books record it in cultivation in Britain in the mid-nineteenth century. It can be recognised by its very broad leaves, similar to lilac, *Syringa vulgaris* L., which gave rise to the specific name.

It is very similar to the widespread *Jasminum elongatum* (Berg.) Willd., but differs in its shorter calyx lobes and corolla tube, and in having a more open inflorescence, not closely subtended by leaf-like bracts. It requires frost-free conditions: hardiness rating G2, USDA zones 9–11.

Jasminum syringifolium Wall. ex G. Don, *Gen. Syst.* 4: 62 (1837). Type: Native of the Burman Empire at Amhurst, on the banks of the Martaban, and at Tavoy. *Wall.* Cat. 2861.
Syn. *J. scandens* Kerr. in Craib, *Fl. Siam* 3: 405 (1939), non (Retz.) Vahl.

DESCRIPTION. Spreading evergreen shrub or woody climber. *Leaves* opposite, simple, ovate to lanceolate, 4–11 × 2–5 cm long, with 3–4 pairs of veins; base rounded to subcordate; apex acute to acuminate; petiole 3–10 mm long, articulate. *Inflorescence* terminal on side shoots, 1–10 cm diameter, 5 to many-flowered, pedicels 4–10 mm. *Bracts* linear, 1–7 mm. *Flowers* white, fragrant, 1–2 cm diameter. *Calyx* tube 1.5–2 mm long, lobes subulate to narrow triangular, 0.5–1.5 mm. *Corolla* tube 8–13 mm, lobes 7–8, oblong–elliptic, 7–11 × 2.5–3 mm. *Berries* black, ellipsoid, 14 × 8 mm.
DISTRIBUTION. Assam, Thailand and Burma.
HABITAT. Low altitude evergreen forests.
FLOWERING TIME. Winter and spring.

Related species:

33a. Jasminum duclouxii (H. Lév.) Rehder, *J. Arnold Arbor.* 15: 307 (1934) is occasionally grown. It is native to south-west China and India and has distinctive leaves, dark green, leathery, narrowly lanceolate to lanceolate, 5.5–18.5 × 1.5–5 cm, with 7–8 or more pairs of veins and long acuminate apex on a stout petiole up to 10 mm. The flowers are borne in winter and spring in clusters up to 15, are very fragrant, white, often pink or reddish-purple outside. The corolla tube widens at the apex and the 4–5 corolla lobes are rounded.

Section ALTERNIFOLIA DC.

Plants shrubby, rarely climbing. Leaves alternate; petioles not articulate. Flowers yellow. Two seeds in each fruit. The species of sect. Alternifolia range from northern and central China through India, Pakistan, Nepal, Afghanistan, Iran and parts of southern Europe with a few outlying species found in Madeira and the Canary Islands and East Africa . The type species is *Jasminum humile*.

34. JASMINUM BIGNONIACEUM

Section Alternifolia DC.

Jasminum bignoniaceum is a shrub with alternate leaves and unscented rich yellow flowers with a funnel-shaped corolla tube widening distinctly to the throat. It was known to be growing at Royal Botanic Gardens, Kew in 1971 from material collected in the Nilghri hills in India, but is rare in cultivation and possibly only available from one or two British or European nurseries. However, there is considerable confusion in gardens between this species and *J. humile* which is found wild in the Sino-Himalayas and which is much more common in cultivation. The flower shape of the rarer *J. bignoniaceum* is distinct with a broader corolla tube and almost funnel-shaped flowers with short, rounded corolla lobes. *J. bignoniaceum* has been grown in California and is thought to have been introduced there from Royal Botanic Gardens, Kew (Green, 1965), as it was included in the Kew seed list from the mid-twentieth century onwards. It has proved to be hardy on a warm sunny or partly shaded wall in very sheltered regions and with its long flowering period would be a useful addition to a mild garden. At Kew it survives against a wall, but the tips of the shoots are often killed by light frosts. Its hardiness rating is H5–G1, USDA zones 8–11.

The leaflet and flower numbers are variable. Plants from Sri Lanka appear consistently to have 5 larger, usually broader leaflets, (10–) 25–30 (–40) × (7–) 10–14 (–17) mm and solitary flowers and may be referred to subsp. *zeylanicum* P. S. Green. This plant has recently been collected from Sri Lanka in 2002 by Bleddyn and Sue Wynn-Jones (BSWJ 9486) of Crûg Farm Plants. Plants from India tend to have 7 or more pairs of smaller and narrower leaflets, (3–) 5–15 (–20) × (1.5–) 3–7 (–10) mm and 3–5 flowers in each inflorescence, and may be referred to subsp. *bignoniaceum*.

Jasminum bignoniaceum Wall. ex G. Don, *Gen. Syst.* 4: 63 (1837). Type: Nilgiri Hills, *Wallich* 2888 (K).
Syn. *J. revolutum* Sims var. *peninsulare* DC.

ILLUSTRATION. Wight, *Spicilegium Neilgherrense* 2, t. 151 (1851), as *J. revolutum*. Fyson, *Fl. Nilgiri & Pulney Hill-tops*, iii. 415 (1920).

DESCRIPTION. Evergreen glabrous shrub or small tree to 2 m or more. *Branches* angled. *Leaves* alternate, usually pinnate with 5–9 leaflets but sometimes trifoliate or simple towards base of shoots; petiole 0.5–1.5 cm. *Leaflets* narrow elliptic to broad ovate to obovate with slightly recurved margin, veins obscure; base cuneate; apex acute; terminal leaflet 1–4 × 0.5–1.5 cm; lateral leaflets 0.5–3 × 0.5–1.3 cm. *Inflorescence* terminal on side shoots, more or less umbellate, 1–5-flowered, pedicels 4–8 mm. *Bracts* filiform, to 5 mm. *Flowers* somewhat pendulous, yellow, not scented, to 10 mm diameter. *Calyx* tube 1.5–2 mm long, lobes to 1 mm. *Corolla* tube 14–17 mm, funnel shaped, expanding from 2 mm wide at base to 6 mm at throat, lobes 5, rounded, sometimes slightly

emarginate, 3–4 × 3–5 mm. *Berries* greenish-white, globose *c.*4 mm.

DISTRIBUTION. Southern India and Sri Lanka.

HABITAT. Mountainous regions; to around 2,000 m.

FLOWERING TIME. Intermittently from spring to early winter.

35. JASMINUM FLORIDUM

Section Alternifolia DC.

Jasminum floridum forms a shrub with alternate leaves and fragrant yellow flowers with long narrow calyx lobes and acute corolla lobes. Like *J. fruticans*, the calyx lobes are long, exceeding the calyx tube, but the shape of the leaflets is distinct being broader in *J. floridum* and with an acute apex. It also resembles *J. humile* and especially the cultivar 'Revolutum' but the leaves never have more than 5 leaflets, usually only 3, and the calyx lobes are considerably longer.

This species was discovered by A. von Bunge in 1831 and the earliest cultivated plants in Europe were known from northern China, but later Augustine Henry found it also in central China. It has long been cultivated in China but the first record in British gardens was the report in *Botanical Register* of 1842 as collected by the Hon. William Fox-Strangeways, later the Earl of Ilchester, and grown in his gardens at Abbotsbury in Dorset.

It is a frost-tolerant plant and easily grown in any soil. Hardiness H3: USDA zones 7–10. The specific epithet, *floridum*, refers to its profuse flowering.

Jasminum floridum Bunge, *Enum. Pl. Chin. Bor.* 42: (1833). Type: N. China, Kantai 1831, *Bunge* (LE).

Syn. *J. giraldii* Diels, *Bot. Jahrb. Syst.* 29: 534 (1900).

    J. subulatum Lindl., *Edward's Bot. Reg.* 28 (Misc.): 57 (1842).

ILLUSTRATIONS. *Curtis's Botanical Magazine* 109: 6719 (1883); Dirr, *Dirr's Trees and Shrubs for Warm Climates*, p. 155 (2002).

DESCRIPTION. Evergreen or semi-evergreen shrub to 3 m. *Branches* 4-angled, green, glabrous or hairy. *Leaves* alternate, usually with 3 leaflets, occasionally 1 or 5, sometimes simple at base of shoots, petiole 2–10 mm long. *Leaflets* ovate to elliptic, rarely obovate, glabrous or hairy, veins obscure; base cuneate to rounded; apex acute, often mucronate; terminal leaflet 0.5–5 × 0.3–2.5 cm; lateral leaflets 0.4–3 × 0.2–1.4 cm. *Inflorescence* terminal on side shoots, usually 6–12-flowered, pedicels usually 10 mm or more. *Bracts* subulate, 3–7 mm. *Flowers* yellow, fragrant. 1.5–2 cm diameter. *Calyx* glabrous or slightly hairy, tube 1–2 mm long, lobes to 2 mm, linear. *Corolla* tube 9–15 × 3–4 mm, lobes 5, ovate or oblong, often acute at apex, 4–8 mm. *Berries* black, globose, 5–10 mm in diameter.

DISTRIBUTION. North and central China, and Japan where it is probably naturalised.

HABITAT. Slopes, valleys, woods, thickets; below 2,000 m.

FLOWERING TIME. Throughout the summer.

Plate 23. *Jasminum bignoniaceum.* Hand-coloured lithograph by Dumphy after a painting by Govindoo, from Wight, *Spicilegium Neilgherrense* 2, t. 151 (1851).

36. JASMINUM FRUTICANS

Section Alternifolia DC.

This species forms a twiggy shrub with alternate leaves and small, fragrant yellow flowers with narrow calyx lobes to 2 mm or more and rounded corolla lobes. It was one of first jasmines to be introduced to cultivation in Europe and the only species found wild there. It is a common shrub in Mediterranean garrigue and other dry shrubby communities.

Though it is mentioned by John Gerard in his Herbal in 1597, it was probably cultivated well before this date. It does not have a strong scent and the flowers are rather small so it has become less common in gardens than some of the more floriferous species. In colder regions, even if the leaves drop, the green stems give some winter interest. It may be grown as a shrub against a wall but would need support. Hardiness rating H4, USDA zones 6–10.

It is close to *Jasminum floridum* but has narrower leaflets and fewer flowers on shorter pedicels.

Jasminum fruticans L., *Sp. Pl.* 1: 7 (1753). Type: Jasminum no. 3, Hort. Cliff. (BM).

ILLUSTRATIONS. *Curtis's Botanical Magazine* 13: 461 (1799); Huxley and Taylor, *Flowers of Greece and the Aegean*, t. 201 (1977); *The Plantsman* 10: 150 (1988); Ellison, *Cultivated Plants of the World*, p. 329 (1995).

DESCRIPTION. Evergreen or semi-evergreen glabrous shrub 1–2 m. *Branches* angled, green. *Leaves* alternate, trifoliolate, occasionally simple; petiole *c.*2–8 mm long. *Leaflets* elliptic or narrowly elliptic with slightly recurved margins, veins obscure, base cuneate; apex rounded; terminal leaflet 0.5–3 × 0.2–1.5 cm; lateral leaflets 0.2–2.5 × 0.1–0.8 cm. *Inflorescence* terminal on side shoots, 1 to 8 flowered; pedicels to 10 mm but usually much less. *Flowers* yellow, somewhat fragrant, to 15 mm diameter. *Calyx* tube 1.5–2.5 mm long, lobes 3–5 mm, linear. *Corolla* tube 8–15 mm, lobes 5, rounded, 5–10 × 3–6 mm, throat hairy within. *Berries* black, globose *c.*7 mm.

Fig. 38. *Jasminum fruticans*.

Plate 24. *Jasminum floridum*. Hand-coloured lithograph by Anne Barnard from *Curtis's Botanical Magazine* t. 6719 (1883).

DISTRIBUTION. Throughout the Mediterranean area, east to northern Iran and adjacent Turkmenistan.

HABITAT. Usually in dry rocky areas, among scrub or in deciduous woodland; 100–1,000 m.

FLOWERING TIME. Summer.

Fig. 39. *Jasminum fruticans*.

Plate 25. *Jasminum fruticans*. Hand-coloured engraving by Sydenham Edwards from *Curtis's Botanical Magazine* t. 461 (1799).

37. JASMINUM HUMILE

Section Alternifolia DC.

This species forms an upright shrub with alternate leaves and yellow flowers in umbels on pedicels usually greater than 1 cm. It has a very wide geographical distribution and as a result shows considerable variation in morphological characteristics and degrees of hardiness. It has been collected by several different expeditions and some collections have, in the past, been given specific status. However, these are now considered to be simple variations of the species.

The earliest plants are thought to have come to Italy by the ancient trade route from south-eastern Asia to Venice. John Tradescant grew the plant in 1656 and during the mid-seventeenth century it could well have been imported, along with orange trees, from Italy which may be the reason for the vernacular name of Italian yellow jasmine. In France, around Paris, it was reported as being used for hedging according to the account in *Botanical Register* 1819, but at this time London gardens were thought to be too damp for it to thrive. This reference also notes that Miller reported that it was grafted onto *Jasminum fruticans* which made the plant hardier. As well as being naturalised in parts of the Mediterranean, it is very widely cultivated in many other parts of the world.

Fruits are formed when plants of more than one clone are grown in proximity. This is easily cultivated and is one of the species which will thrive in a moister soil and dappled shade. It will tolerate considerable frost. Hardiness rating H5, USDA zones 7–10.

Jasminum humile L., *Sp. Pl.* 1: 7 (1753). Type: Herb. Linn. no. 17.6 (LINN).

ILLUSTRATIONS. *Addisonia* 12: t. 412 (1927); Hay & Synge, *Dictionary of Garden Plants*, p. 248 (1969) as *J. revolutum*; *The Plantsman* 10: 150 (1988); *Botanical Register* 5: 350 (1819); Brickell, *The RHS A–Z Encyclopedia of Garden Plants* p. 585 (2003); Ellison, *Cultivated Plants of the World*, p. 329 (1995); Dirr, *Dirr's Trees and Shrubs for Warm Climates*, p. 156 (2002).

DESCRIPTION. Semi-evergreen or evergreen, erect, or sometimes somewhat scrambling, glabrous or hairy shrub of 2–3 m or more. *Branches* angular, green. *Leaves* alternate, dark green, often leathery, pinnate with usually 5 but also from 1–13 leaflets, variable in size from 2–10 cm long; petiole 0.5–3 cm long. *Leaflets* ovate to elliptic to lanceolate, margins recurved, glabrous or hairy below, with 2–4 pairs of obscure veins below; base cuneate to rounded; apex acuminate to rounded; terminal leaflet 1–6 × 0.4–3 cm; lateral leaflets 0.5–4.5 × 0.3–2.2 cm. *Inflorescence* lax, terminal on side shoots, more or less umbellate, 5–25–flowered; pedicels 0.5–3 cm. *Bracts*, if present, linear to 7 mm long. *Flowers* yellow, more or less fragrant, 1.5–2 cm diameter. *Calyx*, tube 2–3 mm long, lobes short or triangular, 0.5–1 mm. *Corolla* tube 10–18 mm, lobes 5, orbicular or ovate, slightly reflexed when fully open, 4–9 × 3–6 cm. *Berries* purple-black, ellipsoid or globose, 4–6 mm.

DISTRIBUTION. The Himalayas from Afghanistan to SW China. Naturalised in parts of the Mediterranean including Italy and Greece.

HABITAT. Woods, thickets; 1,000–3,800 m.

FLOWERING TIME. Summer.

Plate 26. *Jasminum humile* **'Revolutum'**. Hand-coloured engraving from *Curtis's Botanical Magazine* t. 1731 (1815).

The following cultivars, varieties or forms are cultivated:

'Revolutum'

Syn. *J. revolutum* Sims, *Curtis's Bot. Mag.* 42: 1731 (1815).
   J. reevesii Hort.
   J. triumphans Hort.

ILLUSTRATIONS. *Curtis's Botanical Magazine* 42: 1731 (1815); *Botanical Register* 3: 178 (1817); *The Plantsman* 10: 150 (1988).

When compared with the average of the species, this is a larger, erect shrub without a tendency to scramble and with larger leaves, each with 5–7 broader leaflets. The larger flowers reach 2–2.5 cm across, and are slightly fragrant. It can be also distinguished by the slightly protruding stamens which are always longer than the style, indicating that all plants in cultivation are probably of a single clone, vegetatively propagated. It is slightly less hardy than the species. The London nursery, Messrs. Lee and Kennedy were cultivating this plant by 1817. It has received the Royal Horticultural Society's Award of Garden Merit as a good all round garden plant.

Fig. 40. *Jasminum humile* **'Revolutum'**.

37a. forma **wallichianum** (Lindl.) P. S. Green.
Syn. *J. wallichianum* Lindl., *Edwards's Bot. Reg.* 17: t. 1409 (1831).
   J. humile var. *glabrum* (DC.) Kobuski.
ILLUSTRATION. *Edwards's Botanical Register* 17: t. 1409 (1831) as *J. wallichianum*.

This form normally bears 7–13 hairy leaflets but fewer, though scented, flowers. The plant illustrated in 1831 was raised from seeds sent from Nepal by Rev. William Herbert. It mainly occurs in NE Nepal. Many plants in gardens today are the result of seeds collected by the University of North Wales Expedition to Nepal in 1971, (BL&M 241). Even more recently, in 1994, Bleddyn and Sue Wynn-Jones brought material from Lachung, Eastern Sikkim (BSWJ 2559).

37b. forma **farreri** (Gilmour) P. S. Green.
Syn. *J. farreri* Gilmour, *Curtis's Bot. Mag.* 157: 9351 (1934). Type: Upper Burma: Hpimaw hill, just below the DaK bungalow, *Farrer* 867 (E).

ILLUSTRATION. *Curtis's Botanical Magazine* 157: 9351 (1934); *The Plantsman* 10: 150 (1988).

Plate 27. *Jasminum humile* forma *farreri*. Painted by Lilian Snelling for *Curtis's Botanical Magazine*, t. 9351 (1934).

This differs from the species in the downy leaves which are normally trifoliolate, often purple tinged when immature, and the consistently scented flowers to 15 mm. It was first introduced by Reginald Farrer who described it as 'a small graceful bush', from a limestone ridge at 2,286 m in Upper Burma (now Myanmar) in 1919 (*F.* 867). The seeds sent to the Royal Botanic Garden in Edinburgh were tentatively labelled 'very near *J. giraldii* Diels' and the plant was grown under this specific name in British gardens for many years, until John Gilmour corrected it in the article in *Curtis's Botanical Magazine* in 1934.

38. JASMINUM LEPTOPHYLLUM

Section Alternifolia DC.

This dainty evergreen shrub with yellow flowers and alternate, very narrow, undivided leaves is known only from the western edge of the Himalayas in northern Pakistan where it grows in a very remote area with difficult access which may explain why it had remained undiscovered by the botanical world until so recently.

It was collected on a Kew expedition to Palas Valley, northern Pakistan in 1995, where it was recorded that the plant was used locally as a deterrent to fleas by burning the foliage. Studies are being made at Kew of its genetics, and records are maintained of the techniques used for germination and cultivation to build up the stocks. It is closest to *Jasminum floridum* and *J. fruticans* in having the teeth of calyx longer than tube. Some of latter species may have occasional simple leaves, but no plants bear only linear undivided leaves. It could prove to be more or less hardy in cultivation in Britain. So far plants have proved to be tolerant of lime, probably hardy in milder areas and require no special care. At present (2007) a good specimen can be seen just inside the main gates of Kew. The specific name refers to the slender leaves, derived from the Greek leptos meaning small or meagre.

Fig. 41. *Jasminum leptophyllum* at Kew.

Plate 28. *Jasminum leptophyllum*. Painted by Steven Porwol for *Curtis's Botanical Magazine*, t. 384 (2000).

Jasminum leptophyllum Rafiq, *Novon* 6: 295–297 (1996). Type: Pakistan, Kohistan: above Ban-gah, *c.*1,900 m, 1993, *Rafiq* 14091 (RAW).

ILLUSTRATIONS. *Curtis's Botanical Magazine* 17 (1): n.s. t. 384 (2000).

DESCRIPTION. Evergreen or semi-evergreen much branched erect, glabrous shrub to *c.*2 m. *Branches* angled or ribbed, dark grey. *Leaves* alternate, simple, dark green, narrow linear-lanceolate with recurved margins, 15–28 × 1.2–2 mm; base tapering to petiole or stem; apex blunt, apiculate; petiole absent or very short. *Inflorescence* terminal, 1–3 flowered, pedicels 5–10 mm. *Flowers* bright yellow, fragrant, 1–2 cm diameter. *Calyx* tube 1.5–2 mm long, lobes *c.*2–5 mm, linear. *Corolla* tube *c.*15–20 mm, lobes 5, elliptic, *c.*10 × 5 mm. *Berries* black, subglobose *c.*5 mm.

DISTRIBUTION. North–western Pakistan in lower Palas Valley.

HABITAT. Dry rocky outcrops and cliffs; 1,500–2,000 m.

FLOWERING TIME. Mid- to late summer.

39. JASMINUM ODORATISSIMUM

Section Alternifolia DC.

Jasminum odoratissimum is one of the two species native to Madeira and is also widespread in the Canary Islands. It is a yellow-flowered shrub with alternate, usually trifoliolate leaves and can be distinguished from *J. humile* by having the flowers in a corymbose rather than umbellate inflorescence on pedicels less than 1 cm long.

It was cultivated by Philip Miller in 1730 and is thought to have been introduced into European gardens as early as 1656. It is not common in cultivation and requires a warm temperate climate or a cool greenhouse. As an upright shrub, it does not require support. Hardiness G1: USDA zones 9–11.

Plants from East Africa, previously known as *Jasminum goetzeanum* Gilg, are now considered to be a subspecies, *J. odoratissimum* subsp. *goetzeanum* (Gilg) P. S. Green, but are probably not in cultivation. They tend to have 3–5 (rarely 7) leaflets rather than the 1–3 (rarely 5 in cultivation) more typical of subsp. *odoratissimum*.

Jasminum odoratissimum L., *Sp. Pl.* 1: 7 (1753). Type: No. 17.7 in *Herb. Linn.*, (LINN).

ILLUSTRATIONS. *Curtis's Botanical Magazine* 8: 285 (1794); Everett, *New York Botanical Garden Illustrated Encyclopedia of Horticulture* 6: 1849 (1981); Herklots, *Flowering Tropical Climbers* fig. 195 p. 137 (1976).

DESCRIPTION. Evergreen, upright or scrambling, glabrous or rarely hairy, shrub to 6 m. *Branches* round or slightly angled. *Leaves* alternate, usually trifoliolate or with occasionally 5–7 leaflets or simple leaves at base of inflorescence, dark green, glossy, somewhat leathery; sometimes with slightly undulate margins petiole 0.7–2.5 cm. *Leaflets* lanceolate to ovate to broad elliptic, with 2–4 pairs of obscure veins; base cuneate to rounded; apex acute to blunt; terminal leaflet 1–5 × 1–2.5

Plate 29. *Jasminum odoratissimum*. Hand-coloured engraving by Sydenham Edwards for *Curtis's Botanical Magazine*, t. 285 (1794).

Fig. 42. *Jasminum odoratissimum*

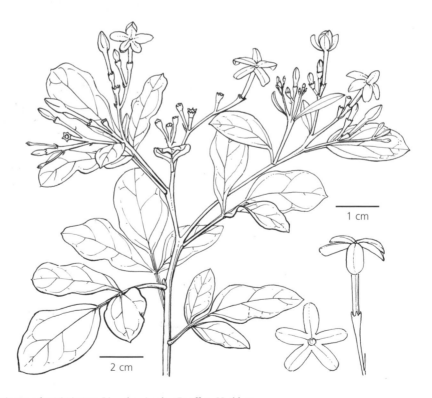

Fig. 43. *Jasminum odoratissimum*. Line drawing by Geoffrey Herklots.

cm; lateral leaflets slightly smaller, 1.3–4 × 0.5–2.5 cm. *Inflorescence* terminal on side shoots, 2–10 cm, 5–25 flowered, pedicels 1–12 mm. *Bracts* linear, 1–2 mm. *Flowers* clear yellow, 10–20 mm across, fragrance variable. *Calyx* tube 1.5–4 mm long, lobes 0.25–2 mm. *Corolla* tube 10–17 mm, lobes 5, broad ovate, 4–10 × 3–5 mm. *Berries* black, glossy, ellipsoid *c*.8–15 mm long.

DISTRIBUTION. Madeira and the Canary Islands.

HABITAT. On cliffs and rocks and in thickets from coast to inland ravines; to about 900 m.

FLOWERING TIME. Intermittently all year round in the wild but mainly from early spring to early summer.

40. JASMINUM PARKERI

Section Alternifolia DC.

This is a hardy dwarf or prostrate shrub with scented yellow flowers and its diminutive size, forming a dense low mound or trailing over rocks, makes it suitable for rock gardens. It appears to withstand somewhat dry conditions and is also suitable for cultivation in a container, making an attractive pot plant in a cool greenhouse. It is hardy to moderate frost, hardiness rating H4, USDA zones 6–10.

Some have considered *Jasminum parkeri* to be a starved form of *J. humile* but it has distinctly lemon-yellow rather than golden yellow flowers and retains its diminutive habit and small flowers in cultivation. In the wild, it has a very restricted distribution in northern India.

Richard Neville Parker (1884–1958), who first introduced the species into cultivation, was in the Indian Forest Service and was chief conservator of forests in the Punjab in 1932.

Jasminum parkeri Dunn, *Bull. Misc. Inform., Kew* 1920: 69 (1920). Type: India, Chamba State, Tiari, Barmaor, on rocks 800 m, 2 July 1919, *R. N. Parker* s.n. (K).

ILLUSTRATIONS. *Gardeners' Chronicle* 158: 509 (1965); Ellison, *Cultivated Plants of the World*, p. 329 (1995); Dirr, *Dirr's Trees and Shrubs for Warm Climates*, p. 159 (2002).

DESCRIPTION. A prostrate or low growing semi-evergreen shrub rarely exceeding 30 cm. *Branches* angled, green. *Leaves* alternate 1–2 cm long, usually with 3–5 leaflets, petiole 1–3 mm long. Leaflets narrow ovate to broadly elliptic with slightly recurved margins; veins obscure; base acute; apex acute to obtuse; terminal leaflet 3.5–5.5 × 2–2.5.mm; lateral leaflets 2.5–4 × 2–2.5 mm. *Inflorescence* terminal on side shoots, 1 to occasionally 3-flowered; pedicels 2 mm. *Bracts* minute. *Flowers* lemon-yellow, fragrant, to 10 mm diameter. *Calyx* with scattered hairs, tube 1.5 mm long, lobes barely 1 mm. *Corolla* tube 12–13 mm, lobes 6, elliptic, 5–6 mm. Berries translucent, greenish-white, globose *c*.4.

DISTRIBUTION. Northern India (NW Himachal Pradesh).

HABITAT. On dry rocks in forests; 800–2,450 m.

FLOWERING TIME. Early summer.

41. JASMINUM SUBHUMILE

Section Alternifolia DC.

Jasminum subhumile is a large shrub or even a small tree with alternate leaves, with usually 3 broad leaflets, and numerous yellow fragrant flowers with very short calyx lobes. The variation in the number of leaflets from one to three on an individual plant has resulted in the synonyms *J. heterophyllum* and *J. diversifolium*. There is also considerable variation in the degree of pubescence, from glabrous to densely hairy. The varietal epithet var. *glabricorymbosum* W. W. Sm. has sometimes been used for the glabrous forms. The species has a geographical distribution similar to that of *J. humile* but it grows at lower altitudes, in the foothills of the southern Himalayas, and is therefore less hardy. It is not dissimilar to *J. humile* but it is a much larger shrub with broader leaves and many more flowers in each inflorescence.

The first record in cultivation is in 1820 as *Jasminum heterophyllum*. It should be treated as a tender plant in all but very warm climates but is worth growing for the numerous fragrant, golden yellow flowers. However it does form a large shrub or even a small tree to 3 metres tall so needs space to reach its full potential. Hardiness G2; USDA zones 9–11.

Jasminum subhumile W. W. Smith, *Notes Roy. Bot. Gard. Edinburgh* 8: 127 (1913). Type: Yunnan, Sha–yang valley, lat. 25°20'N, 1,800 m, moist open situation, April 1910, *Forrest* 5529 (E).
Syn. *J. heterophyllum* Roxb., *Hort. Bengal.* 3 (1814).
> *J. diversifolium* Kobuski, *J. Arnold Arbor.* 20: 404 (1939).

ILLUSTRATION. Phillips and Rix, *Conservatory and Indoor Plants* 2, p. 102 (1997).

DESCRIPTION. Evergreen glabrous or hairy shrub or small tree 1–3 m. Branches slightly angled. *Leaves* alternate, trifoliolate or simple below inflorescence, rather leathery; petiole 0.5–6 cm long. *Leaflets* lanceolate to broad ovate, sometimes hairy below, especially on midrib, with 3–6 pairs of rather obscure veins; base cuneate to rounded; apex acute to long acuminate; terminal leaflet 2–12.5 × 1–6 cm; lateral leaflets 2–10 × 0.7–5 cm. *Inflorescence* terminal on side shoots, 7–12 cm across, 20 – 120-flowered, pedicels 1–20 mm. Bracts linear, 1–5 mm. *Flowers* yellow, fragrant, 10–20 mm diameter. *Calyx* glabrous, tube 1–2 mm long, lobes

Fig. 44. *Jasminum subhumile* var. *glabricorymbosum*, near Dali, Yunnan.

minute. *Corolla* tube 9–12 mm, lobes 4–5, rounded, 4–9 × 3–4 mm. *Berries* black or reddish–black, globose or ellipsoid, 1–1.6 × 0.5–1.6 cm.

DISTRIBUTION. South-west China (Yunnan), Burma, India and Nepal.

HABITAT. Along streams, and on the edges of woods; 700–3,300 m.

FLOWERING TIME. Late spring to early summer.

Section PRIMULINA P. S. Green

The two species of section Primulina P. S. Green are characterised by their yellow flowers and opposite leaves, in contrast to the yellow-flowered species of section Alternifolia, which, as the name suggests, all have alternate leaves. Both species are winter or early spring flowering and have long scrambling rather than twining stems.

42. JASMINUM MESNYI

Section Primulina P. S. Green

Although first described from dried material collected by W. Hancock near Mengtsze in western China and later collected by Augustine Henry from the same locality, it was Ernest Wilson who introduced this plant into western gardens while collecting as an employee of the famous nursery of Messrs. Veitch and Sons. Even then he did not find seed but sent living plants via Hong Kong. Wilson gives an interesting and detailed account of how he brought living plants from Mengtsze to Hong Kong in 1899 and shipped them from there to England in an article in *Gardeners' Chronicle* 21 July 1906. One plant survived, was propagated and, once the stock was built up, it was distributed in 1903. He received 4 rooted cuttings from the wife of the Commissioner of the Customs Station at Mengtsze and also collected four more plants on his way back to Hong Kong.

The name commemorates William Mesny (1842–1919) from Jersey who became a Major-General in the Imperial Chinese army and who found the plant in 1880 during his travels around China.

Jasminum mesnyi can be easily recognised by its scrambling evergreen habit, opposite leaves and non-scented large yellow, usually semi-double, flowers.

It is now extensively cultivated throughout the tropical and subtropical parts of the world, where it is commonly known as the primrose jasmine. It is said to have survived 16 degrees of frost, in dry soil, at Coombe Wood, Veitch's Nursery, but it is advisable to give the plant protection until it is well established, and even then only grow in the most sheltered position on a sunny wall. In colder areas, it might be grown in a cool conservatory or in a container out of doors and brought under shelter in the coldest season to avoid frosts. It is not a strong climber and requires support, to display the rather pendant branches covered with very striking large yellow flowers to their best advantage. It could also be grown to hang down a bank or wall or trained as a standard, pruning it hard after flowering. The trailing branches may root themselves. Seed has not been seen in cultivation, which may be because all plants in cultivation have been derived from a single clone: the crossing of different clones from new introductions may produce fertile fruit.

Initially, *Jasminum mesnyi* was thought to be a form of *J. nudiflorum* because the flowers are variable in size and petal number. Other theories were that this plant showed the true characters of *J. nudiflorum,* which had deteriorated in cultivation in European gardens!

It is, however, probably an ancient cultivar grown in southern China for a very long time. Dr Augustine Henry for example, who collected it in Yunnan, saw it in gardens or hedges around villages in 1882, but did not find it in the forest. There have continued to be disputes as to whether it is simply a cultivated form grown in Chinese gardens, but it is now recognised as a species native to Guizhou, SW Sichuan and Yunnan in China. Its hardiness rating is G1–H5, USDA zones 8–11.

Jasminum mesnyi Hance, *J. Bot.* 20: 37 (1882). Type: China; Guizhou, *Mesny* s.n.(holotype BM).

Syn. *J. primulinum* Hemsl. ex Baker, *Bull. Misc. Inform., Kew* 1895: 109 (1895).

ILLUSTRATIONS. *Curtis's Botanical Magazine* 130: 7981 (1904) (as *J. primulinum*); Noailles & Lancaster, *Mediterranean Plants and Gardens,* 89 (1977); Brickell, *The RHS A–Z Encyclopedia of Garden Plants* 586 (2003); Phillips and Rix *Conservatory and Indoor Plants* 2, p. 102 (1997); *The Plantsman* 10: 152 (1988); Herklots, *Flowering Tropical Climbers,* fig. 196, p. 137 (1976); Dirr, *Dirr's Trees and Shrubs for Warm Climates,* p. 157 (2002); Ellison, *Cultivated Plants of the World,* p. 329 (1995).

DESCRIPTION. Evergreen glabrous scrambling shrub or weak climber to *c*.4 m. *Branches* 4-angled. *Leaves* opposite, trifoliolate, dark green, to 10 cm long; petiole 5–15 mm long. *Leaflets* narrowly ovate or ovate-lanceolate to narrowly elliptic, veins obscure; base cuneate; apex blunt, mucronate; terminal leaflet 25–55 × 5–20 mm; lateral leaflets about half the size. *Bracts* leafy, obovate or

2 cm

Fig. 45. *Jasminum mesnyi*. Line drawing by Geoffrey Herklots.

Plate 30. *Jasminum mesnyi*. Hand-coloured lithograph by Matilda Smith from *Curtis' Botanical Magazine*, t. 7981 (1904).

Fig. 46. *Jasminum mesnyi*

lanceolate, 5–10 mm. *Flowers* solitary on axillary shoots, pedicels 3–8 mm, single or semi-double, bright yellow, orange in throat; not scented, 2–4.5 cm diameter *Calyx*, tube 2–3 mm long, lobes lanceolate, 4–8 × 2–3 mm. *Corolla* tube 9–12 mm, lobes 6–10 or more, obovate to rounded, 15–20 mm long. *Berries* black, ellipsoid *c*.10 × 6 mm but rarely seen in cultivation.

DISTRIBUTION. South and west China.

HABITAT. Ravines, woods; 500–2,600 m.

FLOWERING TIME. Spring.

43. JASMINUM NUDIFLORUM

Section Primulina P. S. Green

Robert Fortune, on one of the first expeditions sponsored by the Royal Horticultural Society, in 1844, brought this species back from gardens and nurseries around Shanghai in western China. Before this it had only been known as a herbarium specimen in the Imperial Russian Chinese Herbarium as *Jasminum angulare*. Initially it was grown as a tender plant in the greenhouse but within four years was flowering out of doors in the Chiswick garden of the Royal Horticultural Society. It has since proved itself to be one of hardiest, most reliable and popular winter flowering plants, with the common name winter jasmine. It is easily recognised as a hardy deciduous shrub with opposite leaves and non-scented yellow flowers.

Jasminum nudiflorum produces bright yellow flowers, often with a more golden yellow reverse and quite variable in size, which stand out so well against the green, leafless stems, anytime from December until March in the northern hemisphere. In extremely cold areas, it may flower later. It is a very adaptable and tough plant that will survive almost any conditions and any soil, but ideally is best grown in a sunny position. As a weak climber with hanging branches, it requires some support if grown against a wall or on a fence. It could also be planted on top of a bank to cascade down.

If it needs to be controlled, cut back the older growth after flowering. Two separate clones are needed for the production of fruit, but trailing branches will often root themselves. It has long been cultivated in China where they 'graft it on the more common kinds about a foot from the ground which improves its appearance' (*Edwards's Botanical Register*, 1846).

The name *nudiflorum* refers to its naked flowers, opening on bare branches before the leaves open. Its hardiness rating is H2, USDA zones 7–10.

Two varieties with variegated leaves are cultivated:

'Aureum' (syn. 'Variegatum'; var. *aureo-variegatum* Hort.) has yellow-blotched or entirely yellow leaves, which may revert to green.

'Mystique' with greyish-green leaflets edged with silvery-white, grown by Japanese nurseries, was introduced from there into cultivation in Britain by Peter Catt in 1992. Other variegated forms with white edged foliage have been listed in the past as 'Argenteum'.

There are also reports of a compact dwarf and slow growing form, 'Nanum', with smaller leaves which may not be in cultivation in Europe, but has been seen recently in gardens in western Sichuan. This is possibly the same as *Jasminum nudiflorum* var. *pulvinatum* (W. W. Sm.) Kobuski, from SW Sichuan, NW Yunnan and SE Xizang. (Rix, pers. comm.).

Fig. 47. *Jasminum nudiflorum* in the garden of a house on Kew Green.

Jasminum nudiflorum Lindl., *Trans. Hort. Soc. London* 1: 153 (1846). Type: Cultivated, London, ex China (Herb. Lindl. CGE).

Syn. *J. angulare* Bunge, *Enum. Pl. China Bor.*: 42 (1833), (non Vahl).

J. sieboldianum Blume, *Mus. Bot.* 1: 280 (1851).

ILLUSTRATIONS. *Curtis's Botanical Magazine* 78: t. 4649 (1852); *The Plantsman* 10: 152 (1988); *Edwards's Botanical Register* 32: 48 (1846); Brickell, *The RHS A-Z Encyclopedia of Garden Plants* p. 586 (2003); Dirr, *Dirr's Trees and Shrubs for Warm Climates*, p. 157 and 158 (2002).

DESCRIPTION. Deciduous glabrous scrambling *shrub* to 5 m. *Branches* 4-angled, green. *Leaves* opposite, trifoliolate to 7–22 × 4–13 mm; petiole 1–1.5 cm. *Leaflets* ovate to elliptic, veins obscure; base cuneate; apex acute to obtuse; terminal leaflet 13–30 × 3–11 mm; lateral leaflets 6–23 × 2–11 mm. *Flowers* solitary, yellow, scentless, 2–2.5 cm diameter, axillary on leafless shoots, pedicels 2–3 mm. *Bracts* leafy, ovate to lanceolate, 3–8 mm long. *Calyx* tube 2–3 mm long, lobes narrow lanceolate, 5–6, 4–7 × 2 mm. *Corolla* tube 1–1.5 cm, lobes 5, broadly ovate, 8–10 mm long. *Berries* black, ovoid or ellipsoid *c*.6 × 4 mm (not usually found in cultivation).

DISTRIBUTION. W China.

HABITAT. Thickets, ravines, slopes; 800–4,500 m.

FLOWERING TIME. Winter and early spring.

Plate 31. *Jasminum nudiflorum*. Hand-coloured lithograph by W. H. Fitch for *Curtis's Botanical Magazine*, t. 4649 (1852).

BIBLIOGRAPHY

Bean, W. J. (1973). *Trees and Shrubs Hardy in the British Isles*. Eighth edition 2: 463–469. John Murray, London.

Burkhill, I. H. (1966). *A Dictionary of Economic Products of the Malay Peninsula* 2. Published on behalf of the Governments of Malaysia and Singapore by the Ministry of Agriculture and Co-operatives, Kuala Lumpa.

Calkin, R. R. & Jellinek, J. S. (1994). *Perfumery: Practice and Principles*. John Wiley, New York.

Cambie, R. C. & Ash, J. (1994). *Fijian Medicinal Plants*. CSIRO, Melbourne.

Chang, M. C. & Green, P. S. (1996). *Jasminum* in *Flora of China* 15: 397–419.

Collins, M. (2000). *Medieval Herbals. The Illustrative Traditions*. British Library and University of Toronto Press, London.

Elliott, W. R. & Jones, D. L. (1990) *Encyclopaedia of Australian Plants Suitable for Cultivation* 5: 474–477.

Gerard, J. (ed.), Thomas Johnson (1633). *The Herball*. Adam Islip, Joice Norton & Richard Whitakers, London.

Goody, J. (1993). *The Culture of Flowers*. Cambridge University Press, Cambridge.

Gray, A. (1846). Analogy between the Flora of Japan and that of the United States. *Amer. J. Sci.,* N. S. 2: 135–136.

Green, P. S. (1961). Studies in the genus *Jasminum* I: section Alternifolia. *Notes Roy. Bot. Gard. Edinburgh* 23: 355–384.

—— (1962). The identity of *Jasminum absimile*. *Baileya* 10: 53–55.

—— (1962). Studies in the genus *Jasminum* II: The species from New Caledonia and the Loyalty Islands. *J. Arnold Arbor.* 43: 109–131.

—— (1963). Studies in the genus *Jasminum* III: The species in North America. *Baileya* 13: 137–172.

—— (1969). Studies in the genus *Jasminum* IV: The so-called New World species. *Kew Bull.* 23: 273–275.

—— (1970). Studies in the genus *Jasminum* V: Plants from two islands off the east coast of Africa. *Kew Bull.* 24: 227–229.

—— (1984). Studies in the genus *Jasminum* VII: *Jasminum laurifolium* as a cultivated plant. *Kew Bull.* 39: 655–656.

—— (1984). Studies in the genus *Jasminum* VIII: A revision of *Jasminum* in Australia. *Allertonia* 3: 403–438.

—— (1985). Studies in the genus *Jasminum* IX: Notes on two Jasmines from South India and Ceylon. *Kew Bull.* 40: 225–230.

—— (1986). Studies in the genus *Jasminum* X: *Jasminum* in Arabia. *Kew Bull.* 41: 413–418.

—— (1987). A long-misapplied name for a Sino-Indian species of *Jasminum*. *Kew Bull.* 42: 437–438.

—— (1992). *Jasminum officinale* 'Inverleith'. *Kew Magazine* 9: 63–67, t. 196.

—— (1993). Jasminum sinense. *Curtis's Botanical Magazine (Kew Magazine)* 10: 113–316, t. 224.

—— (1995). Studies in the genus *Jasminum* (*Oleaceae*) XIV: New species and combinations in *Jasminum*, especially from Thailand. *Kew Bull.* 50: 567–580.

—— (1997). Studies in the genus *Jasminum* (*Oleaceae*) XV: A revision of the pinnate-leaved species of *Jasminum*. *Kew Bull.* 52: 933–947.

—— (1997). In: Cullen, J. *et al.* (eds). *The European Garden Flora* 5: 574–592. Cambridge University Press.

—— (2000). *Jasminum* in *Flora of Thailand* 7(2): 306–340. The Forest Herbarium, Bangkok.

—— (2001). Studies in the genus *Jasminum*, XVII: sections *Trifoliolata* and *Primulina*. *Kew Bull.* 56(4): 903–915.

—— (2003). Synopsis of the *Oleaceae* from the Indian Sub-Continent. *Kew. Bull.* 58: 257–295.

—— (2003). Studies in the genus *Jasminum* (*Oleaceae*). *Kew Bull.* 58: 297.

—— (2005). *Jasminum* in Staples, G. W. & Herbst, D. R., *A Tropical Garden Flora* 438–441.

Guenther, E. (1952). *The Essential Oils*. Van Nostrand, New York.

Guitian, J., Guitian, P. & Medrano, M. (1998). Flora biology of the distylous Mediterranean shrub *Jasminum fruiticans* (Oleaceae). *Nordic J. Bot.* 18(2): 195–201.

Hanelt, P. (ed.) (2001). *Mansfeld's Encyclopedia of Agricultural and Horticultural Crops*. Springer, Berlin and London.

Harborne, J. B. & Baxter, H. (2001). *Chemical Dictionary of Economic Plants*. Wiley, Chichester.

Herklots, G. (1976). *Flowering Tropical Climbers*. Dawson, Folkestone, Kent.

Ide, L. S. (1998). *Hawaiian Lei Making*. Mutual Publishing.

Kiew, R. (1994). Name changes of Malaysian plants: *Begonia wrayi* (Begoniaceae) and *Jasminum aemulum* (Oleaceae). *Malayan Nat. J.* 47(3): 311–317.

—— (1994). Six new species of *Jasminum* (Oleaceae) from Malesia. *Sandakania* 4: 75–81.

Kobuski, C. E. (1932). Synopsis of the Chinsese species of *Jasminum*. *J. Arnold Arbor.* 13: 145–179.

—— (1939). New and noteworthy species of Asiatic *Jasminum*. *J. Arnold Arbor.* 20: 64–72.

—— (1939). Further notes on *Jasminum*. *J. Arnold Arbor.* 20: 403–408.

—— (1959). A revised key to the Chinese species of *Jasminum*. *J. Arnold Arbor.* 40: 385–390.

Kupicha, F. K. (1983). *Jasminum* in E. Launert (ed.), *Flora Zambesiaca* 7 (1): 305–318.

Li, H.-L. (1959). *The Garden Flowers of China*. Ronald Press, New York.

Manandhar, N. P. (2002). *Plants and People of Nepal*. Timber Press, Oregon.

Needham, J. (1986) *Science and Civilisation in China*. 6, part 1: Botany. Cambridge University Press, Cambridge.

Pandy, B. P. (1989). *Sacred Plants of India*. Shree Publishing House, New Delhi.

Parkinson, J. (1639). *Paradisi in Sole Paradisus Terrestris*. Humfrey Lownes & Robert Young, London.

Padua, L. S. de, Bunyappraphatsara, N. & Lemmens, R. H. M. J. (eds) (1999). Plant Resources of South-east Asia: medicinal and poisonous plants. *Prosea* 12 (1).

Rohwer, J. G. (1994). Seed characters in *Jasminum* (Oleaceae): unexpected support for De Candolle's sections. *Bot. Jahrb. Syst.* 116(3): 299–319.

—— (1995). Seed characters in *Jasminum* (Oleaceae). II. Evidence from additional species. *Bot. Jahrb. Syst.* 117(3): 299–315.

—— (1996). Die Frucht-und Samenstrukturen der Oleaceae. *Biblioth. Bot.* 148: 10.

—— (1997). The fruits of *Jasminum mesnyi* (Oleaceae), and the distinction between *Jasminum* and *Menodora*. *Ann. Missouri Bot. Gard.* 84: 848–856.

Salguero, C. P. (2003). *A Thai herbal*. Findhorn Press, Scotland.

Srivastava, S. K. & Kapoor, S. L. 1984. A note on *Jasminum adenophyllum* (Oleaceae), a rare endemic in Meghalaya State (India). *J. Econ.Taxon. Bot.* 5(2): 497–499.

Thompson, J. D. & Dommée, B. (2000). Morph-specific variation in the patterns of stigma height in natural populations of distylous *Jasminum fruticans*. *New Phytol.* 148: 303–314.

Thackston W. M. (1996). Mughal Gradens in Persian Poetry. In: Wescoat, J. L. Jr. & Wolschke-Bulmahn, J. (eds). *Mughal Gardens, Sources, Places, Representations, and Prospects*, pp. 233–257. Dumbarton Oaks, Washington DC.

Valder, P. (1999). *The Garden Plants of China*. Weidenfeld & Nicholson, London.

Verdoon, I. C. (1956). The Oleaceae in Southern Africa. *Bothalia* 6: 549–639.

Wallander, E. & Albert, V. A. (2000) Phylogeny and classification of Oleaceae based on *rps16* and *trnL-F* sequence data. *Amer. J. Bot.* 2000, 87: 1827–1841.

Weiss, E. A. (1997). *Essential Oil Crops*. CABI, Oxfordshire.

PLANT NAME INDEX

Accepted names are in **bold**; synonyms in *italics*. Principal reference page nos. in bold. (Authorities are only included for homonyms).